Nico Vogt

Governing fungal polar cell extension

Nico Vogt

Governing fungal polar cell extension

Analysis of Rho GTPase and NDR kinase signalling in Neurospora crassa

Südwestdeutscher Verlag für Hochschulschriften

Impressum / Imprint

Bibliografische Information der Deutschen Nationalbibliothek: Die Deutsche Nationalbibliothek verzeichnet diese Publikation in der Deutschen Nationalbibliografie; detaillierte bibliografische Daten sind im Internet über http://dnb.d-nb.de abrufbar.

Bibliographic information published by the Deutsche Nationalbibliothek: The Deutsche Nationalbibliothek lists this publication in the Deutsche Nationalbibliografie; detailed bibliographic data are available in the Internet at http://dnb.d-nb.de.

Verlag / Publisher:
Südwestdeutscher Verlag für Hochschulschriften
ist ein Imprint der / is a trademark of
OmniScriptum GmbH & Co. KG
Heinrich-Böcking-Str. 6-8, 66121 Saarbrücken, Deutschland / Germany
Email: info@svh-verlag.de

Herstellung: siehe letzte Seite /
Printed at: see last page
ISBN: 978-3-8381-1004-2

Zugl. / Approved by: Göttingen, GAU, Diss., 2008

Die vorliegende Arbeit wurde in der Arbeitsgruppe von Prof. Dr. Gerhard H. Braus in der Abteilung Molekulare Mikrobiologie des Instituts für Mikrobiologie und Genetik der Georg-August-Universität Göttingen angefertigt.

Teile dieser Arbeit wurden veröffentlicht in:

Seiler, S., Vogt, N., Ziv, C., Gorovits, R. and Yarden, O. (2006). The STE20/germinal center kinase POD6 interacts with the NDR kinase COT1 and is involved in polar tip extension in *Neurospora crassa*. Mol Biol Cell 17, 4080-4092.

Seiler, S., März, S., Ziv, C., Vogt, N., Helmstaedt, K., Cohen, N., Gorovits, R. and Yarden, O. The Ndr kinase COT1, and the MAP kinases MAK1 and MAK2 genetically interact to regulate filamentous growth, hyphal fusion and sexual development in *Neurospora crassa*. Manuscript in revision.

Vogt, N. and Seiler, S. The RHO1 specific GTPase activating protein LRG1 regulates polar tip growth in parallel to Ndr kinase signaling in *Neurospora*. Manuscript in revision.

D7

Referent: Prof. Dr. Gerhard H. Braus

Korreferentin: Prof. Dr. Stefanie Pöggeler

Tag der mündlichen Prüfung:

Für meine Eltern und Geschwister

Acknowledgements

First of all I want to thank Prof. Dr. Gerhard Braus and Dr. Stephan Seiler for supervision, support during the course of my research study and great working conditions. I am particularly thankful that they enable me to present parts of my work on national and international conferences.
I thank Prof. Dr. Stefanie Pöggeler for accepting to co-examine this work.
Special thanks go to Prof. Dr. Oded Yarden, Carmit Ziv and Rena Gorovits from Israel for the great cooperation and a lot of helpful advice and support. I am grateful to PD Jan Faix (Hannover) for the cooperation on the establishment of the Rho GEF assays.
I would like to thank Dr. Andre Zeug and especially Arwed Weigel for their help in microscopy and a lot of interesting discussions. I would like to thank Dr. Stefan Lakämper for the protein expression in SF9 cells.
I am grateful to Prof. Dr. Aaron Neiman (New York) for providing antibodies. I would also like to thank Dr. Angela Hoffman (Portland) for providing cercosporamide. I would like to thank Dr. Sebastian Fettig, Carolyn Rasmussen, Dr. Alf Herzig, Kay Schink, Dr. Anne-Brit Krappmann, Dr. Stefan Eimer and Dr. Damian Brunner for making plasmids available to me and for friendly discussions.
For providing protocols, helpful hints and ideas I would like to thank Prof. Dr. Stephen J. Free (New York), Uta Gey, Prof. Dr. Gero Steinberg and Prof. Dr. Nick Read (Edinburgh). I would also like to thank Dr. Michael Mahlert, Andrea Hlubek and Dr. Nicole Nolting for helpful discussions.
I am grateful to PD Dr. Michael Hoppert for his patiently help in the cooperation on electron microscopy. I also would like to thank Dr. Özgür Bayram for the fruitful cooperation on the velvet project. I would like to thank Dr. Oliver Valerius for the protein identifications and hints on practical aspects of protein work.
I wish to thank Inga von Behrens for her contribution to this work in her diploma thesis and the good collaboration. I would like to thank Seema Singh for her contribution to this work.
Special thanks go to Dr. Elke Schwier for her commitment and support during the course of this work, intensive proofreading of parts of this thesis and a lot of fruitful discussions. I would like to thank Daniela Justa-Schuch for enjoyable discussions in tough working periods. I am grateful to the present members of the *Neurospora* lab Sabine März, Corinna Schmitz and Anne Dettmann for cooperation, proofreading of documents and the pleasant working atmosphere. I would like to thank the present and former members of the department Dr. Lars Fichtner, Christoph Sasse, Jennifer Gerke, Karen Laubinger, Heike Rupprecht, Gaby Heinrich, Dr. Silke Busch, Dr. Tim Köhler, Dr. Katrin Bömeke and Dr. Andrea Grzeganek (nee Pfeil) for being great colleagues. Thanks also to the practical students Melanie Nolte, Christian Timpner, Sonja Oberbeckmann and Immo Röske for their contribution to this work. I would also like to thank all other present and former members of the group for their helpfulness and the pleasant working atmosphere.
I also wish to thank my friends for giving me a lot of general support. I especially want to thank Cindy for motivation in difficult times. My deepest thanks go to my family for their tremendous support.

Table of contents

Acknowledgements I

Table of contents III

Summary 1

Zusammenfassung 2

1 Introduction 3

1.1 Polarity in eukaryotic cells 3

1.2 Polar growth in filamentous fungi 4

1.3 Signal transduction pathways involved in fungal morphogenesis 5

1.3.1 Rho proteins 5

1.3.2 NDR kinases and germinal center kinases in morphogenesis 8

1.4 Aims of this work 10

2 Materials and methods 12

2.1 Strains, media and growth conditions 12

2.2 Plasmid construction 16

2.3 General molecular methods 24

2.3.1 Bioinformatics 24

2.3.2 General cloning procedures 24

2.3.3 Immunological methods 25

2.4 Biochemical methods 25

2.4.1 Protein purification 25

2.4.2 Enzymatic assays 26

2.4.2.1 In vitro assay for Rho GAP activity 26

2.4.2.2 In vitro assay for Rho GEF activity 26

2.5 Microscopy 27

2.5.1 Immunofluorescence 27

2.5.2 GFP fluorescence 27

3 Results 29

3.1 POD6 and LRG1 are essential for hyphal tip extension 29

3.2 pod-6 and lrg-1 deletion mutants show identical phenotypes like conditional mutants 32

3.3 COT1 and POD6 act together to regulate polar tip growth 36

3.4 The function of LRG1 in polar tip elongation 39

3.4.1 LRG1 is a fungal specific protein containing LIM and GAP domains 39

3.4.2 LIM and GAP domains of LRG1 are both essential for growth and septation 42

3.4.3 The LIM domains are required for localizing LRG1 to sites of growth 44

3.4.4 LRG1 is a RHO1 specific GAP 49

3.4.5 LRG1 regulates several output pathways of RHO1 51

3.5 LRG1 acts parallel with the Ndr kinase COT1 in a motor protein dependent manner 55

4 Discussion 60

4.1 Mutations in lrg-1 and pod-6 affect hyphal tip elongation, septation and determination of branching in N. crassa 60

4.2 The germinal center kinase POD6 acts together with COT1 in polar tip extension 60

4.3 The LIM domains are required for the localization of LRG1 61

4.4 LRG1 regulates the activity of several Rho1 effector pathways 63

4.5 Comparison of LRG1 to its homologues in yeasts 66

4.6 The COT1/POD6 complex and LRG1 act in parallel morphogenetic pathways 66

4.7 The influence of RHO cycling on activity 68

5 References 71

6 Supplementary data 84

6.1 Lrg1p and Cbk1p are involved in pseudohyphae formation in Saccharomyces cerevisiae 84

6.2 CDC24 enhances GDP-GTP exchange in vitro 84

6.3 Vector information 86

6.3.1 Cloning plasmid pNV86 86

6.3.2 Expression plasmids pNV87 and pNV88 87

6.3.3 Modification of the multiple cloning sites of pMal-c2x to obtain pNV72 89

6.4 Searching for interacting proteins of LRG1: tandem affinity purification 89

Summary

Polar morphogenesis is required for the function of elongated cell types like neuronal cells, pollen tubes and cells of filamentous fungi. The basal signalling components involved are highly conserved. Defects in polar cell shape can result in developmental disorders or death of the affected cell. In this work two components involved in polar growth were analysed at a molecular level in *Neurospora crassa*. These are LRG1, which is a member of the GTPase activating proteins (GAPs), and the germinal center kinase POD6.

POD6 and LRG1 are proteins essential for hyphal tip elongation. Deletion and temperature sensitive mutants of *pod-6* and *lrg-1* show phenotypic similarities to *cot-1* temperature sensitive mutant in cessation of hyphal elongation and excessive hyperbranching. All three proteins are also involved in determining the size of hyphal compartments.

Complementation analysis revealed that both parts, the N-terminal containing three LIM domains as well as the C-terminal harbouring the Rho GAP domain of LRG1, are required for its function. Genetic evidence and *in vitro* GTPase assays identify LRG1 as a RHO1 specific GAP.

Localisation experiments revealed a partial colocalisation of POD6 and COT1 that depends on the oppositely directed microtubule motor proteins kinesin-1 and dynein. LRG1 shows a similar localisation and is enriched at septae and at hyphal tips. This was observed by immunofluorescence studies with antibodies generated against LRG1 and confirmed in a strain expressing MYC^9::LRG1. In strains expressing GFP::LRG1, the dynamic accumulation of the fusion protein as an apical cap was observed. This localisation depends on the three LIM domains of LRG1, a functional actin cytoskeleton and active growth. Similar to the localisation of COT1 and POD6, LRG1 localisation is influenced by dynein and kinesin-1 and the microtubule cytoskeleton.

LRG1 affects several output pathways of RHO1. Hyposensitivity of *lrg-1(12-20)* to the glucan synthase inhibitor caspofungin and synthetic lethality with a hyperactive b1,3-glucan synthase mutant occurred. Further, suppression by the PKC inhibitors staurosporine and cercosporamide was observed. Hypersensitivity to the actin depolymerising drug latrunculin A and the suppression of defects in *lrg-1* mutant strains by the overexpression of the dominant-negative acting N-terminus of the formin BNI1 indicate an influence on formin mediated actin polymerisation. In contrast, the *cot-1* mutation has no influence regarding these RHO1 effectors.

Taken together, these data suggest that LRG1 functions as a GAP for Rho1 that regulates several effector pathways. A complex of COT1 and POD6 acts in parallel to coordinate apical tip growth.

Zusammenfassung

Die Entwicklung gestreckter Zelltypen wie Nervenzellen, Pollenschläuche und Zellen filamentöser Pilze erfordert polare Gestaltbildung. Die grundlegenden Signalwege dafür sind hochkonserviert. Missbildungen der Zellform können zu Entwicklungsstörungen oder dem Tod der betroffenen Zelle führen. In dieser Arbeit werden Komponenten, die für polares Wachstum von *Neurospora crassa* benötigt werden, auf molekularer Ebene charakterisiert. Dabei handelt es sich zum einen um ein Mitglied der GTPase aktivierenden Proteine, LRG1 und zum anderen um die "germinal center" Kinase POD6.

POD6 und LRG1 sind für das gerichtete Hyphenwachstum unabdingbar. Deletions- und temperatur-sensitive Allele der Gene *cot-1*, *pod-6* und *lrg-1* zeigen phänotypisch Ähnlichkeiten bezüglich des Abbruchs polaren Wachstums, übermäßiger Verzweigungsbildung und der veränderten Größe der Hyphenkompartemente. Komplementationsexperimente belegen, das sowohl der die drei LIM Domänen enthaltende N-Terminus als auch der die Rho-GAP Domäne enthaltende C-Terminus von LRG1 für dessen Funktion benötigt werden. Neben genetischen Hinweisen identifizieren *in-vitro* Untersuchungen LRG1 als Rho1 spezifisches GAP.

COT1, POD6 und LRG1 befinden sich in der Zelle an Stellen aktiven Wachstums. In Immun-Fluoreszenz Untersuchungen wurde eine partielle Ko-Lokalisation von POD6 und COT1 festgestellt. Diese Lokalisierung ist von Mikrotubuli Motorproteinen abhängig. LRG1 zeigt eine ähnliche, an Septen und der Hyphenspitze angereicherte, zelluläre Verteilung. Diese Lokalisation wurde sowohl mit gegen LRG1 gerichteten Antikörpern in Wildtyp-Zellen als auch für MYC^{9}::LRG1 mit Antikörpern gegen das Myc-Epitop gefunden. Ein mit GFP fusioniertes LRG1-Protein findet sich in Abhängigkeit von intakten LIM Domänen und Aktin-Zytoskelett wachstumsabhängig als apikale Kappe, deren Größe mit der Wachstumsgeschwindigkeit korreliert. Ähnlich wie bei POD6 und COT1, wird auch diese Lokalisation wird von den Mikrotubuli Motoren Kinesin und Dynein beeinflusst.

LRG1 beeinflusst die Aktivität verschiedener Effektorproteine von Rho1. Mutanten in *lrg-1* sind unempfindlicher gegenüber dem Glucansynthase-Inhibitor Caspofungin und Doppelmutanten von *lrg-1* mit einer hyperaktiven β-1,3-Glucansynthase Mutante sind nicht lebensfähig. Die PKC-Inhibitoren Staurosporin und Cercosporamid unterdrücken die Wachstumssörungen von *lrg-1* Mutanten, welche zudem empfindlicher gegenüber der Aktin depolymerisierenden Droge Latrunkulin A sind. Zudem wurde eine Unterdrückung der Wachstumsdefekte durch Expression des dominant negativ agierenden N-Terminus des Formins Bni gefunden. Die Untersuchung der Rho1 Wege in einer Mutante der NDR kinase *cot-1* zeigte, dass diese Mutation keinen Einfluss auf die getesteten RHO1 Effektoren hat.

Diese Experimente legen nahe, das LRG1 als spezifisches GAP für Rho1 verschiedene Effektorwege reguliert und das ein paralleler, von Rho1 unabhängiger, Signalweg zum polaren Wachstum erforderlich ist, in welchem COT1 und POD6 agieren.

1 Introduction

1.1 Polarity in eukaryotic cells

Determining and maintaining cell shape is a fundamental prerequisite for proper development of any organism. It is critical for the function of many cell types involved in vectorial processes such as nutrient transport, neuronal signalling, or cell motility. Cell polarization in response to extracellular or intracellular cues follows a common hierarchical schema (Drubin and Nelson, 1996). A spatial cue is required to determine the future site of cell polarization. The origin of this cue depends on intrinsic or external signals, cell type and developmental stage. The cue is serving as a positional mark, interpreted by receptors and the resulting signal subsequently transmitted to downstream signalling networks. Components including plasma membrane proteins, cell wall proteins, extra cellular matrix constituents and cytoskeletal elements function to reinforce the asymmetry induced by the cue. Reorganization of the cytoskeleton and the secretory apparatus follows this initial polarization to maintain cellular polarity.

One of the best-studied model systems at molecular level is the unicellular yeast *Saccharomyces cerevisiae*. In this organism, polarity is coupled to the cell cycle and does not necessarily require external signals, but uses spatial information from the previous cell division. This spatial cue is represented in yeast by the cytoskeleton proteins actin and septins. It is transmitted by GTPases of the Ras (Rsr1p) and Rho (Cdc42p) families, which are organized in a complex with the Cdc42p guanylyl nucleotide exchange factor (GEF) Cdc24p and the scaffold protein Bem1p (Drubin and Nelson, 1996 and references therein). They in turn act to reorganize the cytoskeleton and secretory apparatus, where actin is a critical component for the targeting of the patch.

In other systems like vertebrate epithelial or neuronal cells, the spatial mark depend primarily on surrounding cells or environmental signals and is widely uncoupled from the cell cycle. Determination of the cue requires the integration of multiple signals in order to react in a cell context adequate order. To determine the cue for polarity of epithelial cells, cells require cell-cell and cell-extra cellular Matrix (ECM) adhesion. In animals, this signal integration for establishment of cell polarity is mediated for example in epithelial tissues by the frizzeled/planar cell polarity (PCP) and anterior-posterior patterning signalling pathways (reviewed in Adler, 2002; Axelrod and McNeill, 2002; Mlodzik, 2002; Zallen, 2007). The PCP signal pathway contains six core components and enables the cell to recognize the sites of cell-cell or cell-ECM adhesion and to adapt the cell morphology. Signalling molecules relay spatial information to the downstream components required for polarity establishment, leading to asymmetric organization of the cytoskeleton. In different cell types polarity involves targeted secretion that leads to the deposition of molecules needed for growth, transport or signalling at the chosen site (reviewed in Brennwald and Rossi, 2007).

1.2 Polar growth in filamentous fungi

The fungal kingdom includes an estimated number of 1.5 million species. Fungi are of enormous ecological importance as decomposers of plant material and as symbiotic partners for higher plants. Furthermore, fungi have considerable impact on our economy. They are the most biotechnological useful group of organisms (Gadd, 2007), and cause numerous animal, human and plant diseases (Divon and Fluhr, 2007; Latge, 1999; Ponton *et al.*, 2000; Walsh *et al.*, 2004). The majority of fungi grows exclusively at the apical tip and form filamentous, multicellular hyphae, which are separated by incomplete cross walls (Boyce and Andrianopoulos, 2006; Momany, 2002; Wendland and Walther, 2006). This mode of growth is suggested to be the key to their evolutionary success, which depends mainly on the ability to explore new ecological niches, and to quickly colonize new substrates (Magan, 2007; Morris et al., 2007; Pringle and Taylor, 2002).
Factors that determine and modulate cellular polarity have been the subject of extensive investigations in a variety of fungal model systems (Borkovich *et al.*, 2004; Drubin and Nelson, 1996; Harris, 2006; Nelson, 2003; Wendland, 2001), with the most substantial progress having been made in the yeasts *Saccharomyces cerevisiae* and *Schizosaccharomyces pombe* (Bähler and Peter, 2000; Pruyne and Bretscher, 2000a, b; Pruyne *et al.*, 2004). Apical tip extension is the hallmark of filamentous fungi, and fungal hypha are together with neurons and pollen tubes among the most highly polarized cells (Chilton, 2006; Harris, 2006; Palanivelu and Preuss, 2000; Watabe-Uchida *et al.*, 2006), thus making them attractive models for the analysis of fundamental mechanisms underlying cellular polarity. In addition, the significance of fungi in natural and xenobiotic substrate turnover (Dighton, 2007; Gamauf *et al.*, 2007; Morris *et al.*, 2007), secondary metabolite and protein production (Gadd, 2007; Gloer, 2007; Punt *et al.*, 2002) and their impact as pathogens advice the study of the most fundamental process required for the proliferation of the majority of fungal species – filamentous growth. Understanding this process will have significant implications on our ability to intervene in this process by either inhibiting it in case of detrimental fungi or enhancing it when beneficial growth is desired.
The molecular understanding of fungal morphogenesis is still a major challenge. Phylogenetic analyses and the comparison of *Saccharomyces cerevisiae* morphogenetic data with the limited results from various filamentous asco- and basidiomycetes have established that a core set of

„polarity factors“ are conserved between unicellular and filamentous fungi (reviewed in Borkovich *et al.*, 2004; Harris, 2006; Harris and Momany, 2004; Wendland, 2001). Therefore, the accumulated knowledge of bakers yeast is serving as an invaluable source for comparative morphogenetic studies. Nevertheless, it is becoming increasingly evident that subtle differences in the wiring of these conserved components and the presence of additional proteins that are absent in unicellular fungi result in dramatically different morphogenetic outcomes ranging from budding to true filamentous growth (Boyce *et al.*, 2003, 2005; Li *et al.*, 2006; Malavazi *et al.*, 2006; Rottmann *et al.*, 2003; Seiler and Plamann, 2003).

Differences are also found between unicellular yeasts and filamentous fungi in the organisation of transport processes. Filamentous fungi transport molecules, vesicles and organelles over long distances to enable fast growth, while budding yeast is adapted to nutrient rich environments, where fast intracellular transport over long distances is not required. In *S. cerevisiae* the microtubule skeleton has only an essential function in mitosis. In contrast, in filamentous fungi microtubule based transport is required for many transport processes of cellular components like mitochondria and the endoplasmatic reticulum (ER) (Garcia-Rodriguez *et al.*, 2006) and depends on opposite directed motor proteins, kinesins and dynein (Bruno *et al.*, 1996; Eshel *et al.*, 1993; Ogawa *et al.*, 1987; Seiler *et al.*, 1997; Vale *et al.*, 1985; for reviews see Steinberg, 2000, 2007; Vale, 2003).

1.3 Signal transduction pathways involved in fungal morphogenesis

Polarized growth is a complex multifactorial property, which is coordinated by numerous signals. This signalling network includes GTPases of the Ras super family, the cAMP dependent protein kinase (PKA), the mitogen-activated protein kinase (MAPK) or the nuclear Dbf2-related (NDR) kinase pathways. These pathways are highly conserved and regulate numerous aspects of growth and development like cellular proliferation, differentiation, motility and survival. In fungal systems they are important for maintaining hyphal polarity, development and pathogenicity.

1.3.1 Rho proteins

Rho GTPases are found in all eukaryotic cells and constitute a distinct family within the superfamily of Ras-related small GTPases. The Rho family of GTPases plays a central role in polarized growth in animal and fungal cells (Drubin and Nelson, 1996; Ridley, 1995, 2006). The GTPases act as molecular switches that cycle between an active GTP bound and an inactive GDP bound form and reside in the plasma membrane (Figure 1). Transition between these two forms is achieved through GTPase-activating proteins (GAPs) that stimulate the intrinsic GTPase activity to inactivate the protein and GDP-GTP-exchange factors (GEFs) that catalyse exchange of GDP for GTP to activate the GTPase (Hakoshima *et al.*, 2003; Schmidt and Hall, 2002). In addition to these two regulators, guanine nucleotide dissociation inhibitors (GDIs) block spontaneous activation by forming a complex with the Rho protein and dissociate it from the membrane, resulting in an inactive cytoplasmic pool of GTPases (reviewed in Dovas and Couchman, 2005; Olofsson, 1999).

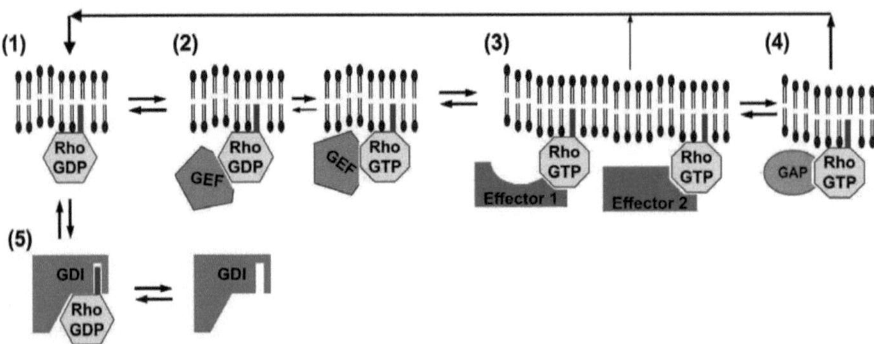

Figure 1: Regulation of Rho GTPase activity.
Rho GTPases are attached to the plasma membrane by their hydrophobic C-terminal prenylation anchor (blue line; 1), where they can be activated. GEF proteins facilitate the GDP to GTP exchange of the GTPase by stabilization of the nucleotide free state (2). In the GTP bound state, Rho proteins interact with and activate different target effectors (red) (3). The intrinsic GAP activity of the GTPase is enhanced by GAPs (green; 4), which leads to the inactive form of the Rho protein (1). GAP proteins specifically influence distinct effectors pathways by their distribution in the cell. Cytosolic GDI proteins (5) bind preferentially to the GDP bound state of Rho proteins and extract them from the membrane by covering the C-terminal prenylation anchor.

Originally, GTPases of the Rho subfamily were described as key regulators of the actin cytoskeleton, but up to date it has become obvious that they influence an amazing variety of cellular processes including cytoskeletal organisation, vesicle transport and transcriptional regulation (for reviews see Etienne-Manneville and Hall, 2002; Jaffe and Hall, 2005; Park and Bi, 2007; Van Aelst and D'Souza-Schorey, 1997). The position of small G-proteins at the bottleneck of many signal transduction pathways does explain the various defects seen in different organisms when these GTPases are misregulated. Twenty-two mammalian genes encoding Rho GTPases have been described (Aspenstrom et al., 2004), whereas *Caenorhabditis elegans* and *Drosophila melanogaster* are predicted to have 10 and 11 genes encoding these proteins, respectively. In the yeast *Saccharomyces cerevisiae* six Rho GTPases named Rho1p to Rho5p and Cdc42p are described (reviewed in Levin, 2005). *N. crassa* also encodes six Rho proteins.

A challenging complication in determining the functions of small G-proteins is the fact that the number of GAPs and GEFs is significantly larger than the number of GTPases. Interestingly, the number of GAPs present in most available genomes exceeds those of the GEFs, suggesting that fine-tuning of the "off-switch" may be important to provide the necessary specificity for the Rho module (Jaffe and Hall, 2005). Although considerable progress has been made in understanding the activation of small GTPases through their GEFs (Garcia *et al.*, 2006b; Gulli and Peter, 2001; Schmidt and Hall, 2002), full activation of a specific GTPase requires not only the coordination of the "on" and "off" switches, but also the shuttling between both the active and inactive state (Barale *et al.*, 2006; Fidyk *et al.*, 2006; Irazoqui *et al.*, 2003; Tu *et al.*, 2002; Vanni *et al.*, 2005; for reviews see also Hall, 2005; Wennerberg and Der, 2004). Therefore, the inactivation through the corresponding GAP is essential for full signalling activity of the small G-protein, but little is known about how Rho proteins are regulated in a spatial and temporal manner. A possible explanation for the high number of GEF and GAP proteins is that they contribute to the spatial and temporal regulation of the Rho GTPases and regulate the activity of the Rho GTPase for specific effectors. Most of these regulatory proteins contain additional domains, which are thought to integrate signals for effective crosstalk between several signal transduction pathways (for reviews see Cote and Vuori, 2007; Tcherkezian and Lamarche-Vane, 2007; Yarwood *et al.*, 2006).

Current fungal research focuses on the characterisation of the various Rho proteins and the analysis of the interplay between the different modules (reviewed in Borkovich *et al.*, 2004; Harris, 2006; Wendland and Philippsen, 2001). Several studies have identified Rho1 as one key regulator of hyphal growth and polarity. *Aspergillus fumigatus* Rho1 has been described as part of the β1,3-glucan synthase complex that localizes to zones of active growth at the hyphal apex (Beauvais *et al.*, 2001). A similar role in maintaining cell wall integrity was suggested for Rho1 of *Ashbya gossypii*, as deletion mutants showed reduced filamentous growth and high rates of cell lysis (Wendland and Philippsen, 2001). *Aspergillus nidulans* RhoA has been implicated in polar growth, branching and cell wall synthesis (Guest *et al.*, 2004). Budding yeast Rho1p as the best characterized representative of the Rho1 family has multiple functions in regulating the two main structural features, the cell wall and the cytoskeleton, of the fungal cell (reviewed in Levin, 2005; Park and Bi, 2007). The organization of the actin cytoskeleton is controlled by the interaction of the activated GTPase with the polarisome component Bni1p, while maintenance of the cell wall integrity is achieved via two independent mechanisms. First, Rho1p activates cell wall synthesis via direct stimulation of the enzyme β1,3-glucan synthase, which catalyzes the polymerization of β1,3-glucan. In addition, it activates through the activation of protein kinase C the Mpk1p/Slt2p MAP kinase pathway that monitors the cell wall integrity. This activation coordinates the transcription of several cell wall specific enzymes. The fungal cell wall composition and structure is different from plant and animal cell compartments and is required for the fitness of pathogenic species. Thus, the cell wall components and factors involved in its regulation are attractive targets for antifungal drug development (for reviews see Bowman and Free, 2006; Latge, 2007).

No data are available for the function of Rho2 and Rho3 in filamentous fungi. The budding yeast protein Rho2p appears to function in a partially redundant manner with Rho1p (Helliwell *et al.*, 1998; Madaule *et al.*, 1987; Ozaki *et al.*, 1996), while Rho3p is important for coordinated polarization of the actin cytoskeleton and the secretory apparatus and is active in concert with Rho4p in *S*. cerevisiae (Adamo *et al.*, 1999; Doignon *et al.*, 1999; Imai *et al.*, 1996; Kagami *et al.*, 1997; Matsui and Toh-E, 1992a, b; Roumanie *et al.*, 2002). Rho4 type GTPases are evolutionary highly divergent and cluster in two evolutionary group. The archae- and euascomycetes Rho4 have been shown to be required for the formation of septae (Nakano *et al.*, 2003; Rasmussen and Glass, 2005, 2007; Santos *et al.*, 2003). Furthermore, Rho4 is involved in cell wall integrity (CWI) signalling in *Schizosaccharomyces pombe* (Santos *et al.*, 2003).

The main focus of research in filamentous fungi currently lies on the function of Cdc42 and the closely related GTPase Rac, primarily because a bona fide Rac is absent from hemiascomycete genomes (Boyce *et al.*, 2001, 2003, 2005; Chen and Dickman, 2004; Mahlert *et al.*, 2006; Virag *et al.*, 2007; Weinzierl *et al.*, 2002). Rac and Cdc42 homologues in the dimorphic fungi *Penicillium marneffei* or *Ustilago maydis* have overlapping, but distinct roles during polarized growth and development. Cdc42 function is required for vegetative hyphal polarity and yeast cell growth in *P. marneffei*, but not for polarization of conidiophores. In contrast, Rac is required for polarized growth during hyphal and asexual development, but not during the yeast phase. Due to lethality of the double, but not the single mutants, both *U. maydis* proteins share at least one essential function. In addition Rac is necessary for the switch from budding to hyphal growth, while Cdc42 regulates cell separation. In *Aspergillus nidulans*, Rac is involved in asexual development, but is not essential for polarity, while Cdc42 is more important for polar growth and lateral branching. Double mutants are also not viable, indicating that they share at least one common function.

1.3.2 NDR kinases and germinal center kinases in morphogenesis

In recent years, protein kinases of the NDR Ser/Thr protein kinase family have emerged as being important for normal cell differentiation and polar morphogenesis in various organisms, yet their specific functions are still elusive (Hergovich *et al.*, 2006; Tamaskovic *et al.*, 2003). In *Drosophila melanogaster*, the NDR kinases Tricornered and Warts are required for control of the extent and direction of cell proliferation as well as for neuronal morphogenesis (Emoto *et al.*, 2004; Geng *et al.*, 2000; Justice *et al.*, 1995; Xu *et al.*, 1995). The *Caenorhabditis elegans* NDR kinase SAX1 regulates aspects of neuronal cell shape and has been proposed to be involved in cell spreading, neurite initiation, and dendritic tiling (Gallegos and Bargmann, 2004; Zallen *et al.*, 2000). Verde and coworkers (1998) have shown that the fission yeast NDR kinase gene *orb6* is required to maintain cell polarity during interphase. The budding yeast NDR kinase Cbk1p is involved in cell separation and modulates cell shape (Bidlingmaier *et al.*, 2001; Racki *et al.*, 2000). A number of large-scale screens have identified several proteins that interact with Cbk1p (Du and Novick, 2002; Ho *et al.*, 2002; Ito *et al.*, 2001), establishing the idea that Cbk1p and other interacting proteins may represent the core components of a conserved complex required for polarized morphogenesis. Further work in fission and budding yeasts as well as in animal cells has resulted in an emerging network, which includes the NDR kinase and its binding partner and activator MOB2 as well as a furry-like scaffolding protein. The NDR kinase is further regulated through a Ste20 type kinase that interacts with MO25 (Hergovich *et al.*, 2006; Kanai *et al.*, 2005; Nelson *et al.*, 2003; Stegert *et al.*, 2005). The founding member of the NDR family, the *Neurospora crassa* kinase COT1, is required for hyphal tip elongation (Collinge *et al.*, 1978; Collinge and Trinci, 1974; Yarden *et al.*, 1992), and temperature-sensitive *cot-1* strains cease hyphal elongation after being shifted to restrictive temperature. This is accompanied by a massive induction of new hyphal tip formation, creating the typical barbed-wired morphology of *cot-1* cells. A similar branching and growth-termination phenotype has been observed in neuronal cells of *sax-1, trc, furry* and *hippo* mutants in *C. elegans* and *D. melanogaster* (Emoto *et al.*, 2004; Emoto *et al.*, 2006; Geng *et al.*, 2000; Zallen *et al.*, 2000), suggesting an evolutionarily conserved function of NDR kinase complexes in the formation of branched cellular structures. An important link between the cytoskeleton and function and COT1 activity has been established by the analysis of *cot-1* suppressor mutants, which are defective in the microtubule-dependent motor protein complex dynein/dynactin (Bruno *et al.*, 1996; Plamann *et al.*, 1994), but the underlying molecular mechanisms are unclear.

Another large emerging group of kinases that have been implicated in various signalling pathways are the Ste20 kinases (Bokoch, 2003; Dan *et al.*, 2001). Originally defined by *S. cerevisiae* Ste20p, an upstream kinase of the mitogen-activated protein kinase pathway, the Ste20 group of kinases is divided into the p21-activated (PAK) kinases and several germinal center kinase (GCK) subfamilies. A C-terminal kinase domain defines the true PAKs and an N-terminally located Cdc42/Rac interacting/binding (CRIB) motif mediates binding of the kinase to the small G-protein and the subsequent activation of the kinase. PAKs were originally characterized as the primary downstream effectors of Rac/Cdc42-type GTPases. The GCK subgroups of PAK kinases differ from the PAK subgroup. The GCK kinase domain is located N-terminally, GC kinases lack the typical CRIB domain, and their noncatalytic domains are highly variable. In contrast to the true PAK subgroup, the function of the GCKs is much less defined, but they have been implicated in stress response, proliferation and apoptosis (Bokoch, 2003; Dan *et al.*, 2001). GCK are suggested to act as activators of NDR kinases although only Cdc15p has been documented as an authentic upstream kinase so far (Hergovich *et al.*, 2006 and references therein).

1.4 Aims of this work

Despite the relevance of a polar growing tip for filamentous fungi, the key components that are specifically required for tip extension and branch-point specification are poorly understood, with COT1 being the best-characterized protein at the starting point of this thesis (Gorovits *et al.*, 1999; Gorovits *et al.*, 2000; Gorovits and Yarden, 2003; Terenzi and Reissig, 1967; Yarden *et al.*, 1992). In a large-scale screen conducted to isolate conditional mutants defective in hyphal morphogenesis Seiler and Plamann (2003) identified several mutants specifically defective in hyphal tip extension. The screen led to the identification of the GC kinase *pod-6* (polarity defective-6), the β1,3-glucan synthase (*gs-1*) and *lrg-1* (LIM and Rho-GAP domain containing protein). The morphological similarities of these mutants, indicated by excessive hyperbranching and cessation of hyphal elongation, suggested a common mechanistic basis for the corresponding proteins during tip elongation in *N. crassa* (Figure 2).

In order to achieve a better molecular understanding of hyphal tip elongation, the functional relationship between COT1, POD6, GS1 and LRG1 was characterised in the course of this work.

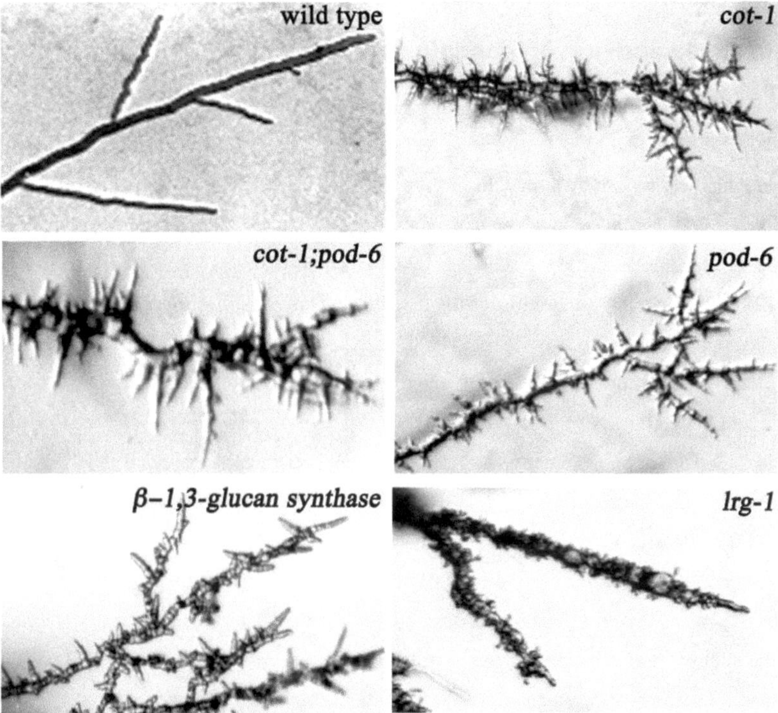

Figure 2: Polarity defective *cot*-like strains
Mutants impaired in hyphal tip elongation and branching were grown at permissive conditions and shifted to restrictive temperature for 8 h to visualise the cessation of tip extension and subsequent hyperbranching. It is important to note that all mutants were still able to establish polarity in germinating conidia and during branch formation at restrictive conditions, but newly emerged tips ceased growth with a pointed tip. (Modified from S. Seiler)

POD6 is a member of the GCK family of Ste20 kinases. In the first part of this work the relation between COT1 and POD6 is analysed. The main part focuses on LRG1, starting with phenotypical characterisation and a domain analysis. The specificity of the LRG1 GAP function for the six RHO GTPases of *N. crassa* is determined and defines LRG1 as a RHO1 specific GAP. The regulation of RHO1 specific effector pathways is analysed, with respect to GS1 and PKC activity and the influences on BNI1-dependent actin organization. The localization of LRG1, its relation to the LIM domains, active growth and the influence of the oppositely directed microtubule-dependent motor proteins dynein and kinesin1 are further investigated. Finally, connections and differences between LRG1 and COT1 signalling in coordinating apical tip growth are analysed.

2 Materials and methods

2.1 Strains, media and growth conditions

General procedures and media used in the handling of *N. crassa* have been described (Davis and DeSerres, 1970) or are available at the Fungal Genetic Stock Center (www.fgsc.net). *N. crassa* strains used in this work are listed in table 1. Strains were grown in either liquid or solid (supplemented with 2% agar) Vogel's minimal media with 2% (w/v) sucrose, unless otherwise stated. When required, H_2O_2 (7 mM), NaCl (0.5–1.2 M), sorbitol (0.5–1.75 M), staurosporine (5 μM) or KT5720 (50 μM), all purchased from Sigma (St. Louis, USA), or latrunculin A (gradient from 0 to 2 μM), purchased from Calbiochem (Merck KGaA, Darmstadt, Germany) were added to the growth medium. A crude preparation of cercosporamide was obtained from Dr. Angela Hoffman (Portland State University, Portland, USA). OmniTray single well plates (Thermo Fisher Scientific, Wiesbaden, Germany) were used as gradient plates. These plates contained solid Vogel's minimal media with 1% sucrose (w/v) and 1% sorbose (w/v) to restrict the radial growth rate (Mishra and Tatum, 1972; Taft *et al.*, 1991). Inhibitors were added to the medium at 50°C, the plates were slanted during the solidification of the agar, then overlaid with an equal volume of the same medium lacking additives in horizontal position and incubated for one day to allow equal diffusion of the additive. For crossings, plates with 0.1% glucose, 2% corn meal agar (Sigma, St. Louis, USA) and if necessary supplemented with 0.1 μg/ml panthothenic acid were used.

DNA transformation of *N. crassa* spheroplasts was carried out as described (Vollmer and Yanofsky, 1986). To select for transformants, the concentrations of hygromycin B and nourseothricin were adjusted to 200 μg/ml and 30 μg/ml, respectively.

To generate deletion mutant strains, homologous recombination events in *N. crassa* were forced by split marker transformation. Primers used in this study are summarized in table 3 and plasmids used in this study are summarized in table 4. The 3' region of *pod-6* and *lrg-1* with a part of the nourseothricin resistance were amplified from the plasmids pNV46 for *lrg-1* and pNV79 for *pod-6* by using the primers NV_nat3 in combination with NV_KOlrg3r2 for *lrg-1* or NV_KOpod3r2 for *pod-6*, respectively. The 5' region of *pod-6* and *lrg-1* with an overlapping part of the nourseothricin resistance were amplified from the same plasmids with primers NV_nat4 in combination with NV_KOlrg5f2 for *lrg-1* and NV_KOpod5f2 for *pod-6*. The overlapping amplicons of the deletion constructs were transformed in the heterokaryotic strain HP1 (Nargang *et al.*, 1995). Strains containing the Δ*lrg-1* or Δ*pod-6* nucleus were verified by the following procedure: First, phenotypic analysis of transformants grown on 400 μM p-fluorophenylalanine (fpa) and 200 μg/ml histidine or

5 μg/ml benomyl and 200 μg/ml pantothenic acid was performed and revealed several candidates with the expected phenotype. Integration of the deletion construct at the *lrg-1* and *pod-6* loci in the genome was verified by PCR. Homologous integration was verified with primers that anneal outside the sequence of the deletion construct in combination with primers that anneal within the resistance cassette. To confirm the 5' genomic integration, NV_KOlrg5_test1 and NV_KOpod5_test1 were used for *lrg-1* and *pod-6*, respectively, in combination with NV_nat4. To confirm the 3' genomic integration NV_nat3 in combination with NV_KOlrg3_test1 and NV_KOpod3_test1 were used for *lrg-1* and *pod-6*, respectively. Finally, complementation of the growth defects of potential *Δlrg-1* and *Δpod-6* strains with a 6 kb genomic *Sac*II fragment containing *lrg-1* from plasmid pNV3 (table 4) or with the *pod-6* carrying cosmid X1F7 verified that the deletion was the reason for the observed phenotype.

A *cot-1(1)* strain transformed with pCZ218 resulted in *myc::cot-1*, which expresses a MYC^6-tagged version of COT1, and was obtained from Carmit Ziv (The Hebrew University of Jerusalem, Rehovot, Israel). Further *N. crassa* strains obtained from S. Seiler or the FGSC are listed in table 1. Double mutants were obtained from crosses of the respective single mutants. Plasmids, which were transformed to obtain new strains, are also mentioned in table 1. *GFP::LRG1* and *GFP::LRG1** were obtained by transformation of the *Δlrg-1* strain with pNV23 or pNV24, respectively. Several rounds of growth on selective medium separated the nuclei of these strains. *GFP::LRG1* and *GFP::LRG1** strains contain therefore only one type of nucleus and conidia do not separate in different phenotypes. They further require panthothenic acid for growth. *GFP::LRG1;nkin* and *GFP::LRG1;ro-1* are derived from crosses of *GFP::LRG1* with *nkin* or *ro-1*, respectively. Other plasmid containing strains of *lrg-1* are transformants derived from *lrg-1(12-20)*.

Table 1: *Neurospora crassa* strains used in this study
(EC): ectopically integrated; in case of ectopic integrations, the transformed plasmid is mentioned. FGSC: Fungal Genetics Stock Center (University of Missouri, Kansas City, USA)

Strain	Genotype	Plasmid	Source
cot-1(1)	*cot-1(C102t)*		FGSC #4066
cot-1(1);Bni1$^{1\text{-}824}$	*cot-1(C102t) bni-1$^{1\text{-}824}$(EC)*	pNV83	this study
cot-1(1);Bni1$^{1029\text{-}1817}$	*cot-1(C102t) bni-1$^{1029\text{-}1817}$(EC)*	pNV84	this study
cot-1(1);gs-1(8-6)	*cot-1(C102t) gs-1(8-6)*		this study
cot-1(1);ro-1	*cot-1(C102t) ro-1(B15)*		Seiler *et al.* (2006)
cot-1(1);ro-10	*cot-1(C102t) ro-10(AR7)*		Seiler *et al.* (2006)
cot-1(1);ro-3	*cot-1(C102t) ro-3(R2354)*		Seiler *et al.* (2006)
GFP::LRG1	*ben^R $his\text{-}3^+$ fpa^S $pan\text{-}2^-$ nat^R:: lrg-1Δ gfp::lrg-1::hph(EC)*	pNV23	this study

Table 1. ***Neurospora crassa*** **strains used in this study (continued)**

Strain	Genotype	Plasmid	Source
*GFP::LRG1**	*ben^R his-3$^+$ fpa^S pan-2$^-$ nat^R:: lrg-1Δ gfp::lrg-1(1-1279; LIM1*, LIM2*, LIM3*)::hph(EC)*	pNV24	this study
GFP::LRG1;nkin	*Nkin(RIP-1) nat:: lrg-1Δ gfp::lrg-1::hph(EC)*		this study
GFP::LRG1;ro-1	*ro-1(B15) natR:: lrg-1Δ gfp::lrg-1::hph (EC)*		this study
gs-1(8-6)	*gs-1(8-6)*		Seiler and Plamann (2003)
gs-1(8-6);lrg-1(12-20)	*gs-1(8-6) lrg-1(Y926H)*		this study
gul-1	*gul-1(CA1)*		FGSC #803
gul-1;cot-1(1)	*gul-1(CA1) cot-1(C102t)*		FGSC #1962
gul-1;lrg-1(12-20)	*gul-1(CA1) lrg-1(Y926H)*		this study
HP1	*ben^R his-3$^+$ fpa^S pan-2$^-$ + ben^S his-3$^-$ fpa^R pan-2$^+$*		Nargang *et al.* (1995)
*LIM1**	*lrg-1(Y926H) lrg-1(C121S, C124S, C98L, C101S) ::hph(EC)*	pNV13	this study
*LIM2**	*lrg-1(Y926H) lrg-1(H185V, C188S, C162S, C165A) ::hph(EC)*	pNV14	this study
*LIM3**	*lrg-1(Y926H) lrg-1(C492S, C495S, C469G, C472S) ::hph(EC)*	pNV15	this study
*LRG1**	*lrg-1(Y926H) lrg-1(1-1279; LIM1*, LIM2*, LIM3*)::hph(EC)*	pNV18	this study
lrg-1(12-20)	*lrg-1(Y926H)*		Seiler and Plamann (2003)
lrg-1; Bni1$^{1-824}$	*lrg-1(Y926H) bni-1$^{1-824}$(EC)*	pNV83	this study
lrg-1;Bni1$^{1029-1817}$	*lrg-1(Y926H) bni-1$^{1029-1817}$(EC)*	pNV84	this study
lrg-1;ro-1	*lrg-1(Y926H) ro-1(B15)*		this study
lrg-1;ro-10	*lrg-1(Y926H) ro-10(AR7)*		this study
lrg-1;ro-3	*lrg-1(Y926H) ro-3(R2354)*		this study
LRG1$^{1-847}$	*lrg-1(Y926H) lrg-1(1-847)::hph(EC)*	pNV7	this study
LRG1$^{781-1279}$	*lrg-1(Y926H) lrg-1(781-1279)::hph(EC)*	pNV6	this study
LRG1$^{781-1279}$::MYC	*lrg-1(Y926H) lrg-1(781-1279)::myc^9::hph(EC)*	pNV21	this study
LRG1^{K910A}	*lrg-1(Y926H) lrg-1(K910A)::hph(EC)*	pNV9	this study
LRG1^{R847L}	*lrg-1(Y926H) lrg-1(R847L)::hph(EC)*	pNV82	this study
myc::cot-1	*cot-1(C102t) myc::cot-1(EC)*		Carmit Ziv, Rehovot, Israel
MYC::LRG1$^{1-847}$	*lrg-1(Y926H) myc^9::lrg-1(1-847)::hph(EC)*	pNV22	this study
nkin	*nkin(RIP-1)*		Seiler *et al.* (1997)
nkin;lrg-1	*lrg-1(Y926H) nkin(RIP-1)*		this study
pod-6(31-21)	*pod-6(31-21)*		Seiler and Plamann (2003)
pod-6(31-21); cot-1(1)	*pod-6(31-21) cot-1(C102t)*		Seiler *et al.* (2006)
ro-1	*ro-1(B15)*		FGSC #146
ro-10	*ro-10(AR7)*		FGSC #3619
ro-3	*ro-3(R2354)*		FGSC #3
TAP::LRG1	*lrg-1(Y926H) tap::lrg-1(EC)*	pNV75	this study
wild type	*74-OR23-1A*		FGSC #987
Δlrg-1	*ben^R his-3$^+$ fpa^S pan-2$^-$ nat:: lrg-1Δ + ben^S his-3$^-$ fpa^R pan-2$^+$ lrg-1$^+$*		this study

Table 1. ***Neurospora crassa*** **strains used in this study (continued)**

Δmak-1	*hph::mak-1D*		FGSC #11321
Δpod-6	*ben*^{R} *his-3*$^+$ *fpa*^{S} *pan-2*$^-$ *nat:: pod-6Δ* + *ben*^{S} *his-3*$^-$ *fpa*^{R} *pan-2*$^+$ *lrg-1*$^+$		this study

For molecular cloning, *E. coli* DH5α (Invitrogen, Carlsbad, USA) and *Saccharomyces cerevisiae* strains Σ1278b, FY2, and BY4742 (Brachmann *et al.*, 1998) were used (table 2). For pseudohyphal growth studies, BY4743, Y33937 and Y22051 were used.

Table 2: Yeast strains used in this study
[a] European *Saccharomyces cerevisiae* Archive for Functional analysis, Johann Wolfgang Goethe-University, Frankfurt, Germany

Strain	Mating type	Genotype	Reference
Σ1278b	MATα	ura3-52 his3Δ::hisG trp1Δ::hisG	(Bechet *et al.*, 1970)
FY2	MATα	ura3-52	(Brachmann *et al.*, 1998)
BY4742	MATα	his3Δ leu2Δ0 lys2Δ0 ura3Δ0	(Brachmann *et al.*, 1998)
BY4743	MATa/MATα	his3Δ1/his3Δ1 leu2Δ0/leu2Δ0 met15Δ0/MET15 lys2Δ0/LYS2 ura3Δ0/ura3Δ0	(Brachmann *et al.*, 1998)
Y33937	MATa/MATα	his3Δ1/his3Δ1 leu2Δ0/leu2Δ0 met15Δ0/MET15 lys2Δ0/LYS2 ura3Δ0/ura3Δ0 YDL240w::kanMX4/YDL240w::kanMX4	EUROSCARF[a]
Y22051	MATa/MATα	his3Δ1/his3Δ1 leu2Δ0/leu2Δ0 met15Δ0/MET15 lys2Δ0/LYS2 ura3Δ0/ura3Δ0 YNL161w::kanMX4/YNL161w	EUROSCARF[a]

Saccharomyces cerevisiae strains were grown at 30°C under non-selective conditions on YEPD (2% pepton, 1% yeast extract, 2% glucose) or for pseudohyphal formation on SLAD medium (0,17% yeast nitrogen base w/o amino acids, w/o ammonium sulfate, 2% glucose, 2% agar complemented with 2 μg/ml histidine, 10 μg/ml leucine and 2 μg/ml uracil).

For protein expression, *Pichia pastoris* GS115 (*his4*) (Invitrogen, Carlsbad, USA), *E. coli* BL21 (DE3), *E. coli* Rosetta2 (DE3) (both Novagen, belongs now to Merck KGaA, Germany) were used. SF9 insect cells were used in cooperation with Stefan Lakämper (Georg-August-Universität Göttingen, Germany). *E. coli* strains were grown on lauria bertani (LB) medium (0.5% sodium chloride, 0.5% yeast extract, 1% pepton) if needed supplemented with 100 μg/ml ampicillin, 50 μg/ml chloramphenicol, 20 μg/ml tetracycline, 50 μg/ml gentamycine or 50 μg/ml kanamycin (all from Sigma, St. Louis, USA) or LBLS medium (0.25% sodium chloride, 0.5% yeast extract, 1% pepton) for zeocin selection supplemented with 50 μg/ml zeocin (Invitrogen, Carlsbad, USA). For growth on solid media, 1.5% agar was added to the medium. For protein expression in *E. coli* Rosetta2 (DE3) LB-rich medium (0.5% sodium chloride, 0.8% yeast extract, 1.8% pepton, 20 mM PIPES, pH 6.7, 2% glucose) was used.

2.2 Plasmid construction

Primers and plasmids used in this study are summarized in table 3 and table 4, respectively.

Table 3: Primers used in this study

Name	Sequence
31-21 3' BamHI	5'-CGG GAT CCT ACG ACA GGC AGC GGC T-3'
31-21 5' NcoI	5'-CAT CCC ATG GCA CCT TTG GAA AAG CTG TTT GA-3'
Hyg3'ApaI	5'-TTT GGG CCC TGA GCG TAT TGG GTG TTA CGG AGC-3'
Hyg3'XbaI	5'-ACC TCT AGA CAA GTG TAC CTG TGC ATT C-3'
Hyg5'ApaI	5'-AAT GGG CCC TGA CAC AGC TCA ATA AGG CTA GCC-3'
Hyg5'XbaI	5'-AGG TCT AGA GTC GGT GAG TTC CTT TC-3'
LRG-3NcoI	5'-TGG CCA TGG CGA TGT CGG TCT TGG ACC CTT G-3'
LRG-5SacI	5'-CTA TGA GCT CCC AAG TAC AGG CGA CAC-3'
NV_CDC24_5	5'-CCC GTC GAC GCC GGG TTC TAA GAT GAC CCA TC-3'
NV_CDC24_6	5'-CCC GCG GCC GCT CAA GCA ACT GGG GCC GCT TGC-3'
NV_Gap1	5'-CGC GAT ATC ATG GCT CCA ATG GTG GAA GG-3'
NV_Gap2	5'-CGC GAA TTC TCA CGT CCC GGG CCC CAC GC-3'
NV_GAPmut_f	5'-GTT CAG GTG GCT GCC TTG CTA GCA CGA TAC CTC CGA GAG CTG C-3'
NV_GAPmut_r	5'-GCA GCT CTC GGA GGT ATC GTG CTA GCA AGG CAG CCA CCT GAA C-3'
NV_GFP1	5'-CCC GAT ATC TGA GCA AGG GCG AGG AGC TG-3'
NV_GFP2	5'-TTT GAT ATC TGT ACA GCT CGT CCA TGC CGA G-3'
NV_GFP5	5'-CCC GGA TCC GGT ACC TTA CTT GTA CAG CTC GTC CAT GC-3'
NV_GFP10	5'-CAG ATC TAT GGG CGA GCA GAA GCT GAT CTC CGA GGA AGA CCT CAA CGG TGT GAG CAA GGG CGA GGA GCT G-3'
NV_gpd1	5'-GGG TTT CGA ACT ACA TCA AGG GTC CAA GAC CGA CAT CGA GGC TCT GTA CAG TGA CCG GTG-3'
NV_gpd2	5'-TAC TAC CAA CAG TCT CCT CGG GTC CTT GCG GGG GCA GTC AGC TCT GTA CAG TGA CCG GTG-3'
NV_GST2	5'-CAG ATC TAT GGG ACG AGG CAG CCA TCA CCA TCA CCA TCA CAA CAC-3'
NV_GST3	5'-CAG ATC TGG CGC CCT GAA AAT AAA GAT TCT C-3'
NV_KOlrg3_test1	5'-CAA AGT CTG CCT GTA AGT CTG-3'
NV_KOlrg3f	5'-GCG AGC GGC AGG CGC TCT ACA TGA GCA TGC CCT GCC CCT GAG CAC TGC ATG AGC TCC TAC-3'
NV_KOlrg3r	5'-GGA ATT GTG AGC GGA TAA CAA TTT CAC ACA GGA AAC AGC CAC CTG GCT CGA CTA AAA CTG-3'
NV_KOlrg3r2	5'-CCA CCT GGC TCG ACT AAA ACT G-3'
NV_KOlrg5_test1	5'-TCT TCT TTC TGC TGT CCT GTC-3'
NV_KOlrg5f	5'-TAA GTT GCG TAA CGC CAG GGT TTT CCC AGT CAC GAC GCA TCG AAT TTG GAA AAT TGG GAC-3'
NV_KOlrg5f2	5'-CAT CGA ATT TGG AAA ATT GGG AC-3'

Table 3: Primers used in this study (continued)

Name	Sequence
NV_KOlrg5r	5'-CTC CGC ATG CCA GAA AGA GTC ACC GGT CAC TGT ACA GAG CCT CGA TGT CGG TCT TGG ACC-3'
NV_KOpod3_test1	5'-GCA AGG CTA AGC TTA CTG ACT G-3'
NV_KOpod3f	5'-GCG AGC GGC AGG CGC TCT ACA TGA GCA TGC CCT GCC CCT GAG GGA GGT AGG GTC TTG-3'
NV_KOpod3r	5'-TGG AAT TGT GAG CGG ATA ACA ATT TCA CAC AGG AAA CAG CGC ATG TGC GGG TGG GTA ATG-3'
NV_KOpod3r2	5'-GCA TGT GCG GGT GGG TAA TG-3'
NV_KOpod5_test1	5'-GTT GGA ATT GCC GGG TAC AAC TC-3'
NV_KOpod5f	5'-ATT AAG TTG CGT AAC GCC AGG GTT TTC CCA GTC ACG ACG CTA CAG CAC TTG TGA TGG TGC-3'
NV_KOpod5f2	5'-GCT ACA GCA CTT GTG ATG GTG C-3'
NV_KOpod5r	5'-CTC CGC ATG CCA GAA AGA GTC ACC GGT CAC TGT ACA GAG CTG ACT GCC CCC GCA AGG ACC-3'
NV_LIM1mut_f	5'-GAT GGG ACC TTT CAT TTG GAT TCA TTC AAG AGT CGC GTG AGT GCC TG-3'
NV_LIM1mut_r	5'-CAG GCA CTC ACG CGA CTC TTG AAT GAA TCC AAA TGA AAG GTC CCA TC-3'
NV_LIM1mut2_f	5'-CTA GCG GAC AAG TGC GAG TGC TTA AGA AAG GTG GTG AAC CTT TGA CGG G-3'
NV_LIM1mut2_r	5'-CCC GTC AAA GGT TCA CCA CCT TTC TTA AGC ACT CGC ACT TGT CCG CTA G-3'
NV_LIM2mut_f	5'-CGC AAA TAC CAC GTC GAC GTC TTT ACC AGC TCG CTT TGC CCG ACT GTC-3'
NV_LIM2mut_r	5'-GAC AGT CGG GCA AAG CGA GCT GGT AAA GAC GTC GAC GTG GTA TTT GCG-3'
NV_LIM2mut2_f	5'-CGC AGG TTA GGC CTG CTG AGT TAC CAG GCC GGC GGT GCT CTT CGG GGC-3'
NV_LIM2mut2_r	5'-GCC CCG AAG AGC ACC GCC GGC CTG GTA ACT CAG CAG GCC TAA CCT GCG-3'
NV_LIM3mut_f	5'-GAC AAG AGG TGG CAT ATC ACG TCT GTC AAC TCC TCA CGT TGC CAG AAA GAA C-3'
NV_LIM3mut_r	5'-GTT CTT TCT GGC AAC GTG AGG AGT TGA CAG ACG TGA TAT GCC ACC TCT TGT C-3'
NV_LIM3mut2_f	5'-GCC TCG GAC TCG GAC TCG GGT ACC CTC TCG AAA AAG CCC ATT GAG GAC GAG-3'
NV_LIM3mut2_r	5'-CTC GTC CTC AAT GGG CTT TTT CGA GAG GGT ACC CGA GTC CGA GTC CGA GGC-3'
NV_link7	5'-CCC AAG CTT GCG GCC GCC GGT ACC GGT CGA CTT GGC CAT GGA TCC GAA TTC TGA AAT CC-3'
NV_lrg3	5'-CAA GCG GCC GCG TCA CCA TGG CTC CAA TGG TG-3'
NV_lrg14	5'-CCC CTC GAG GTC CGG GCC TGG TTC AGG C-3'
NV_lrg18	5'-AGC GGC CGC TCA TAA TTC ATC GGG AAC CAA ACA CAT CTC TTC AAT ATT TGC-3'
NV_LRG20	5'-CTT TCG GTT GAA GGC GTC TTC CTT AAG AAC GGC AAT ATC AAG AAG C-3'
NV_LRG21	5'-GCT TCT TGA TAT TGC CGT TCT TAA GGA AGA CGC CTT CAA CCG AAA G-3'
NV_MBP1	5'-GGT CTC GCT GAA GTC GGT AAG-3'

Table 3: Primers used in this study (continued)

Name	Sequence
NV_myc1	5'-CGG AAT TCC GGC TGG GGC AGG CCA AAC AAT GGG GTG ACT ACT GGC ACT GCA TCT TCT AGA GGT GAA CAA AAG TTG-3'
NV_myc2	5'-TCC CCG CGG CTA CCC GTC AGA TCT GTT CAA G-3'
NV_myc3	5'-GGG CCC GGA TGG CCA TCC CGT CAG ATC TGT TCA AG-3'
NV_myc4	5'-GGG CCC CCA TGG GAT CTT CTA GAG GTG AAC AAA AGG GGC CCC CAT GGG ATC TTC TAG AGG TGA ACA AAA G-3'
NV_nat1	5'-ACC CCA TGG CCA TGA CCA CTC TTG ACG AC-3'
NV_nat2	5'-AGG GAA TTC TCA GGG GCA GGG CAT GC-3'
NV_nat3	5'-ACT CTT GAC GAC ACG GCT TAC-3'
NV_nat4	5'-TAC GCG TGG ATC GCC GGT G-3'
NV_nat5	5'-CCA AAA CAA TAT GGT AGG TGA GGT AGG AGC TCA TGC AGT GCT CAG GGG CAG GGC ATG C-3'
NV_nat6	5'-TTC TGA GAC AAA TAA CAT CCC GTT ACA AGA CCC TAC CTC CCT CAG GGG CAG GGC ATG CTC-3'
NV_Rho1_2	5'-CAA GCG GCC GCT CTG CTG AAC TCC GCC GAA AG-3'
NV_Rho1_3	5'-CTT GCG GCC GCT TAG ACC GAG CTC TTG CAG AGG-3'
NV_Rho2_3	5'-CAA GCG GCC GCT CAT AGA ATC ACA CAG CAC CC-3'
NV_Rho2_4	5'-CAA CCA TGG GCG GCC GCG CAT CAG GCA GCC CTC AG-3'
NV_Rho3_2	5'-CAA GCG GCC GCC CTT GCG GAC TCG GAG GGT C-3'
NV_Rho3_3	5'-CAA GCG GCC GCT TAC ATG ACC ACG CAC TTC G-3'
NV_Rho4_2	5'-CAA GCG GCC GCA CCG AGG GCC CGG CCT AC-3'
NV_Rho4_3	5'-CAA GCG GCC GCC TCA CAT CAT ACC ACA CTT TC-3'
NV_SepA_1a	5'-CAG ATC TTC CTC CCA CGA CAA GAA TGG GAG-3'
NV_SepA_5	5'-CAC TAG TCA TCC TGT TGT TTC TTT ACT TTC TTC AG-3'
NV_SepA_6	5'-CCA GAT CTG GCC CTC CAC CTC CAC CAC C-3'
NV_SepA_7	5'-CCA CTA GTC ACC CTT GGT GGC ATT GGA GGC-3'
NV_tap_N_1	5'-GTT ATC CAT GGC AGG CCT TGC G-3'
NV_tap_n_r	5'-GGG GAT ATC CTA GGG CGA ATT GGG TAC CGG G-3'
Rho1_DN1	5'-GTC TAC GTC CCT ACC GTT TTC ATT AAT TAC GTC GCC GAT GT-3'
Rho1_DN2	5'-AAC CTC GAC ATC GGC GAC GTA ATT AAT GAA AAC GGT AGG GA-3'
Rho1_GV1	5'-CGT CAT CGT TGG CGA CGT CGC CTG CGG CAA GAC C-3'
Rho1_GV2	5'-GGT CTT GCC GCA GGC GAC GTC GCC AAC GAT GAC G-3'
SSe_CDC42_Not3	5'-GAT GCG GCC GCT CAC AGA ATC AAG CAC TTC TTG TCC-3'
SSe_CDC42_Sal5	5'-ACG CGT CGA CCG TGA CGG GAA CTA TCA AGT GCG-3'
SSe_Rac_Not3	5'-GAT GCG GCC GCT TAG AGG ATA GTG CAC TTG GAC TTC-3'
SSe_Rac_Sal5	5'-ACG CGT CGA CCG CTG CTA TCG GAG GCG TGC AGT C-3'

Table 4: Plasmids used in this study.

FGSC: Fungal Genetics Stock Center (University of Missouri, Kansas City, USA)

Name	Description	Source
BAC NC20 K10	*bni* containing Bacmid	FGSC
pAG25	*nat*^{R} (yeast)	Goldstein and McCusker (1999)
pBluescript SK+	cloning vector	Stratagene, La Jolla, USA
pBS1761	encodes TAP tag	Puig *et al.* (2001)
pCZ218	*cot-1::myc*	Carmit Ziv, Rehovot, Israel
pETM-30	expression plasmid, for GST fusions	protein expression facility, Heidelberg, Germany
pFastBac Dual	expression plasmid	Invitrogen, Carlsbad, USA

Table 4: Plasmids used in this study (continued).

Name	Description	Source
pJet1	subcloning vector	Fermentas, Vilnius, LT
pMal-c2x	expression plasmid, for MalE fusions	NEB, Ipswich, USA
pMP6	hygromycin resistance gene under control of modified CPC promoter (*hph*)	M. Plamann, USA
pNV1	pBluescript SK+; P_{gpdA}::*nat*^{R}	this study
pNV2	pBluescript SK+; *lrg-1* coding 6 kb genomic region	this study
pNV3	pBluescript SK+; *lrg-1; hph*	this study
pNV4	pBluescript SK+; *lrg-1*$^{aa781-1279}$	this study
pNV5	pJet1; *gfp*	this study
pNV6	pBluescript SK+; *lrg-1*$^{aa781-1279}$, *hph*	this study
pNV7	pBluescript SK+; *lrg-1*$^{aa1-847}$, *hph*	this study
pNV8	pBluescript SK+; *lrg-1*^{K910A}	this study
pNV9	pBluescript SK+; *lrg-1*^{K910A}, *hph*	this study
pNV10	pBluescript SK+; *lrg-1*$^{C121S, C124S, C98L, C101S}$ (LIM1*)	this study
pNV11	pBluescript SK+; *lrg-1*$^{H185V, C188S, C162S, C165A}$ (LIM2*)	this study
pNV12	pBluescript SK+; *lrg-1*$^{C492S, C495S, C469G, C472S}$ (LIM3*)	this study
pNV13	pBluescript SK+; *lrg-1*$^{C121S, C124S, C98L, C101S}$ (LIM1*)*; hph*	this study
pNV14	pBluescript SK+; *lrg-1*$^{H185V, C188S, C162S, C165A}$ (LIM2*)*; hph*	this study
pNV15	pBluescript SK+; *lrg-1*$^{C492S, C495S, C469G, C472S}$ (LIM3*)*; hph*	this study
pNV16	pBluescript SK+; *lrg-1* coding 6 kb genomic region; reverse	this study
pNV17	pBluescript SK+; *lrg-1*$^{C121S, C124S, C98L, C101S, H185V, C188S, C162S, C165A, C492S, C495S, C469G, C472S}$ (LRG1*)	this study
pNV18	pBluescript SK+; *lrg-1*$^{C121S, C124S, C98L, C101S, H185V, C188S, C162S, C165A, C492S, C495S, C469G, C472S}$ (LRG1*)*; hph*	this study
pNV19	pBluescript SK+; *lrg-1::myc*^{9}*; hph*	this study
pNV20	pBluescript SK+; *myc*^{9}*::lrg-1; hph*	this study
pNV21	pBluescript SK+; *lrg-1*$^{aa781-1279}$*::myc*^{9}*; hph*	this study
pNV22	pBluescript SK+; *myc*^{9}*::lrg-1*$^{aa1-847}$*; hph*	this study
pNV23	pBluescript SK+; *GFP::lrg-1; hph*	this study
pNV24	pBluescript SK+; *GFP::lrg-1*$^{C121S, C124S, C98L, C101S, H185V, C188S, C162S, C165A, C492S, C495S, C469G, C472S}$(LRG1*)*; hph*	this study
pNV25	pMal-c2x; *lrg-1*$^{aa781-848}$	this study
pNV26	pETM-30; *lrg-1*$^{aa781-848}$	this study
pNV27	pETM-30; *lrg-1*$^{aa650-1035}$	this study
pNV28	pETM-30; *rho-1* cDNA	this study
pNV29	pETM-30; *rho-2* cDNA	this study
pNV30	pETM-30; *rho-3* cDNA	this study
pNV31	pETM-30; *rho-4* cDNA	this study
pNV32	pETM-30; *rac* cDNA	this study
pNV33	pETM-30; *CDC42* cDNA	this study
pNV34	pNV80; *rho-1* coding sequence	S. Seiler
pNV35	pNV80; *rho-2* coding sequence	S. Seiler
pNV36	pNV80; *rho-3* coding sequence	S. Seiler
pNV37	pNV80; *rho-4* coding sequence	S. Seiler
pNV38	pNV80; *rac* coding sequence	S. Seiler
pNV39	pNV80; *CDC42* coding sequence	S. Seiler
pNV40	pNV80; *rho-1*^{G15V} coding sequence	this study

Table 4: Plasmids used in this study (continued).

Name	Description	Source
pNV41	pNV80; *rho-1*^{E41I} coding sequence	this study
pNV46	pRS416; *5'lrg-1::nat*^{R}*::3'lrg-1*	this study
pNV47	pNV80; include *myc*^{3} tag	S. Seiler
pNV63	pNV1; deletion of *Nco*I site	this study
pNV70	pNV72; *lrg-1*$^{aa650\text{-}1035}$	this study
pNV72	pMal-c2x; changed multiple cloning site	this study
pNV74	pQE60; *pod-6*$^{aa421\text{-}675}$	this study
pNV75	pBluescript SK+; *TAP::lrg-1; hph*	this study
pNV79	pRS416; *5'pod-6::nat*^{R}*::3'pod-6*	this study
pNV80	pBluescript SK+; *Hyg*^{R}, P_{CPC}	Seiler *et al.* (2006)
pNV81	pBluescript SK+; *lrg-1*^{R847L}	this study
pNV82	pBluescript SK+; *lrg-1*^{R847L}; *hph*	this study
pNV83	pNV47; *bni*$^{1\text{-}824}$	this study
pNV84	pNV47; *bni*$^{1029\text{-}1817}$	this study
pNV85	pETM-30; *cdc24*$^{204\text{-}544}$	this study
pNV86	pPicholi-C; P_{cup1} substituted by *nat*^{R} and *Eco47*IR-*gfp*	this study
pNV87	pFastBac Dual; RGS-His^{6}-GST coding sequence under PH promotor control	this study
pNV88	pNV87: P_{P10}*::gfp*	this study
pPicholi-C	expression plasmid	MoBiTec GmbH, Göttingen, Germany
pQE60	expression plasmid	Qiagen, Hilden, Germany
pRS316-myc^{9}	9 fold *myc* epitope containing yeast plasmid	AB Krappmann
pRS416	Yeast vector used for recombinations	Sikorski and Hieter (1989)
pSM1	P_{gpdA}*::gfp*	Pöggeler *et al.* (2003)
XIF7	*pod-6* containing cosmid	Orbach (1984)

To generate the GAP and LIM domain deletion constructs, a 6 kb genomic *Sac*II fragment containing *lrg-1* coding region and 1.5 kb 5' and 1 kb 3' regions, was inserted in pBluescript SK+ (Stratagene, La Jolla, USA) resulting in plasmids pNV2 and, in reverse orientation, pNV16. The hygromycin B resistance cassette (*hph*), amplified from plasmid pMP6 as template with the primers Hyg5'XbaI and Hyg3'XbaI was inserted into the unique *Xba*I site of pNV16 resulting in the full-length *lrg-1* complementation plasmid pNV3. LRG1$^{1\text{-}847}$ containing plasmid pNV7 was generated by cutting pNV16 with *Apa*I to remove the C-terminal domain of LRG1 to insert the *hph* cassette, which was amplified from plasmid pMP6 with the primers Hyg5'ApaI and Hyg3'ApaI. For generation of pNV6, the *lrg-1* promoter was amplified using oligonucleotides LRG-5SacI and LRG-3NcoI containing *Sac*I and *Nco*I sites, respectively. The *lrg-1* promoter was inserted together with an *Nco*I/*Sac*II fragment from pNV16 containing the GAP domain-encoding region of LRG1 (aa 781-1279) into *Sac*I/*Sac*II digested pBluescript SK+ to obtain pNV4. The *hph* cassette from pNV3 was ligated into the unique *Xba*I site to obtain pNV6. Point mutations in the three LIM domains were generated with the QuickChange® Site-Directed Mutagenesis Kit (Stratagene, La

Jolla, USA) according to the manual using pNV2 as template. Oligonucleotides NV_LIM1mut_f, NV_LIM1mut_r, NV_LIM1mut2_f and NV_LIM1mut2_r were used for mutations in LIM domain 1 to obtain pNV10, NV_LIM2mut_f, NV_LIM2mut_r, NV_LIM2mut2_f and NV_LIM2mut2_r were used for mutations in LIM domain 2 to obtain pNV11 and NV_LIM3mut_f, NV_LIM3mut_r, NV_LIM3mut2_f and NV_LIM3mut2_r were used for mutations in LIM domain 3 to obtain pNV12. The zinc coordinating cysteine residues were substituted mainly by serines that have similar characteristics, but lack the zinc-binding capability. The coordinating histidine was substituted by valine (LIM1: C121S, C124S, C98L, C101S; LIM2: H185V, C188S, C162S, C165A; LIM3: C492S, C495S, C469G, C472S). The hygromycin B resistance cassette from pNV3 was inserted into the unique *Xba*I site of pNV10, pNV11 and pNV12 to obtain pNV13, pNV14 and pNV15, respectively. The triple LIM domain mutation of plasmid pNV17 was generated by multiple mutagenesis PCRs to obtain all 8 mutations of the first and second LIM domain. An *MscI/SacI* fragment from the resulting vector containing mutated LIM1,2*, and a *SacI/MluI* fragment of LIM3* from pNV12 was ligated into pNV16 digested with *Msc*I/*Mlu*I to obtain pNV17. The *hph* cassette from pNV7 harbouring *Bsp120*I (*Apa*I isoschizomere) restriction sites was subsequently inserted into the unique *Not*I site of the vector to obtain pNV18. The K910A exchange in the GAP domain encoded from the plasmid pNV8 was generated by PCR with pNV2 as template and the primers NV_GAPmut_f and NV_GAPmut_r. The hygromycin B resistance cassette was inserted as *Bsp120*I fragment from pNV7 into the unique *Not*I site of pNV8 resulting in pNV9. The R847L mutation within the GAP domain was generated with pNV2 as template and primers NV_LRG20 and NV_LRG21 to obtain pNV81. The *hph* cassette from pNV3 was cloned into the *Xba*I site of plasmid pNV81 resulting in pNV82. N-terminal MYC^{9}-tagged versions of LRG1 and LRG1$^{1\text{-}847}$ were encoded from the plasmids pNV20 and pNV22. To construct these plasmids, a 9-fold MYC-tag was amplified from pRS316-myc^{9} as template using primers NV_myc4 and NV_myc3. The 9-fold MYC-tag was inserted into the unique *Msc*I site one base pair downstream of the LRG1 start codon of plasmid pNV16. The *hph* cassette from pNV7 was subsequently inserted via *Bsp120*I/*Not*I to obtain pNV20 encoding full length MYC^{9}::LRG1 and via *Apa*I to obtain pNV22 encoding MYC^{9}::LRG1$^{1\text{-}847}$. At the C-terminus of LRG1 and LRG1$^{781\text{-}1279}$, the MYC^{9}-tag was inserted via the *Eco*RI and *Sac*II sites of pNV16 and pNV4, respectively. The MYC^{9}-tag was amplified from pRS316-myc^{9} as template using primers NV_myc1 and NV_myc2. The *hph* cassette from pNV7 was subsequently inserted via *Bsp120*I/*Not*I resulting in pNV19 encoding LRG1::MYC^{9} and pNV21 encoding LRG1$^{781\text{-}1279}$::MYC^{9}. To express GFP tagged LRG1 in *N. crassa*, the GFP coding region was amplified from pSM1 (Pöggeler *et al.*, 2003) using oligonucleotides NV_GFP1 and NV_GFP2, and subsequently ligated via *Eco*RV into *Msc*I digested pNV16 and pNV17. The hygromycin cassette from pNV3 was inserted into the *Xba*I site of each plasmid to obtain pNV23 encoding GFP::LRG1 and pNV24 encoding GFP::LRG1*. To construct the resistance cassette for the generation of *Δlrg-1* and *Δpod-6* strains, the *Aspergillus nidulans gpdA* promoter was obtained as an 888 base pair *Sac*I/*Nco*I fragment from pSM1 (Pöggeler *et al.*, 2003). The 581 base pair *nat*^{R} from pAG25 (Goldstein and McCusker, 1999) was amplified with the primers NV_nat1 and NV_nat2 containing *Nco*I and *Eco*RI sites, respectively. The *gpdA* promoter and the *nat*^{R} fragments were ligated into pBluescript SK+ (Stratagene, La Jolla, USA) via *Sac*I/*Eco*RI, resulting in the plasmid pNV1. The *lrg-1* deletion cassette of plasmid pNV46 and the *pod-6* deletion cassette of plasmid pNV79 were obtained by plasmid gap repair in *S. cerevisiae* (Orr-Weaver and Szostak, 1983). The yeast vector pRS416 (Sikorski and Hieter, 1989) was cut with *Xba*I and *Xho*I. The following PCR fragments were used for the recombination. To generate the *nat*^{R} amplicon with pNV1 as template, for the *lrg-1* deletion cassette the primers NV_gpd1 and NV_nat5, and for the *pod-6* deletion cassette the primers NV_gpd2 and NV_nat6 were used in PCR reactions. The 5' and 3' flanking regions of *pod-6* and *lrg*-1 were amplified from cosmid pXIF7 from the Orbach/Sachs library (Orbach, 1984) or from genomic DNA, respectively. For this, primers NV_KOlrg5f, NV_KOlrg5r, NV_KOlrg3f and NV_KOlrg3r for *lrg-1* flanking regions and NV_KOpod5f, NV_KOpod5r, NV_KOpod3f and NV_KOpod3r for *pod-6* flanking regions were used.

For LRG1 antigen expression in *E. coli*, an *lrg-1*$^{aa781\text{-}848}$ fragment was amplified with the primers NV_Gap1 and NV_Gap2 containing the restriction sites *Eco*32I and *Eco*RI, respectively. This fragment was inserted into pMal-c2x (New England Biolabs (NEB), Ipswich, USA) using the *Xmn*I and *Eco*RI sites resulting in pNV25. The*lrg-1*$^{aa781\text{-}848}$ fragment was amplified with NV_lrg3 and NV_Gap2 containing the restriction sites *Nco*I and *Eco*RI, respectively and ligated into pETM-30 (EMBL protein expression facility, Heidelberg, Germany) resulting in pNV26. cDNAs of the six Rho GTPases were generated by reverse transcription with RevertAid M-MuLV Reverse Transcriptase (Fermentas, Vilnius, LT) from *N. crassa* mRNA prepared with PolyATtract (Promega, USA). The RHO coding sequences were amplified by PCR using the following primers amplifying the genes from ATG to stop codon. NV_Rho1_2 and NV_Rho1_3 were used for amplification of *rho-1*, NV_Rho2_4 and NV_Rho2_3 were used for amplification of *rho-2*, NV_Rho3_2 and NV_Rho3_3 were used for amplification of *rho-3*, NV_Rho4_2 and NV_Rho4_3 were used for amplification of *rho-4*, SSe_CDC42_Sal5 and SSe_CDC42_Not3 were used for amplification of *cdc42* and SSe_Rac_Sal5 and SSe_Rac_Not3 were used for amplification of *rac*.

Sequences were inserted via primer-based restriction sites (*Nco*I - *Not*I for *rho-2*, *Not*I - *Not*I for *rho-1*, *rho-3* and *rho-4* or *Sal*I - *Not*I for *cdc42* and *rac*) into pETM-30. First, the *rho-2* sequence was inserted via *Nco*I and *Not*I to obtain the plasmid pNV29. The *rho-2* sequence of pNV29 was cut out with *Not*I and the coding sequences of RHO1, RHO3 and RHO4 were inserted to obtain the plasmids pNV28, pNV30 and pNV31, respectively. The coding sequences of RAC and CDC42 were inserted as *Sal*I/*Not*I fragments into pNV72 (see supplementary data), cut out with enzymes *Nco*I/*Not*I and ligated into pETM-30 cut with the same enzymes to obtain pNV32 encoding RAC and pNV33 encoding CDC42. For generation of the POD6 antigen the coding sequence of amino acids 421–675 was amplified with the primers 31-21 3' BamHI and 31-21 5' NcoI and ligated into pQE60 (Qiagen, Hilden, Germany), resulting in pNV74. The GAP domain of LRG1 consists of amino acids 650-1035. The coding region of this domain was amplified with primers NV_lrg14 and NV_lrg18 and inserted via the generated *Xho*I and *Not*I-sites into pNV32 cut with *Sal*I and *Not*I to obtain pNV27 and into pNV72 resulting in the plasmid pNV70. The GEF domain of CDC24 consists of amino acids 204-544. The coding region *cdc24*$^{aa204\text{-}544}$ was amplified with NV_CDC24_5 and NV_CDC24_6 and ligated into pNV32 via *Sal*I and *Not*I to obtain pNV85. The overexpression constructs encoding *N. crassa* RHO proteins were amplified from genomic DNA and ligated into pNV80. In this vector, the expression of genes is under the control of a constitutive active, modified *cpc-1* promoter derived from plasmid pMP6. The plasmid pNV80 harbours *Bgl*II and *Spe*I as unique cloning sites (Seiler *et al.*, 2006). RHO protein coding regions and the dominant constructs of the formin *bni-1* (NCU01431) were amplified with primers containing the *Bgl*II and *Spe*I restriction sites by S. Seiler (primers for *rho* genes are not listed), to obtain the plasmids pNV34 (for RHO1 expression), pNV35 (for RHO2 expression), pNV36 (for RHO3 expression), pNV37 (for RHO4 expression), pNV38 (for RAC expression) and pNV39 (for CDC42 expression). Plasmid pNV34 was used as template in a mutagenesis to obtain the dominant-active (G15V) as well as dominant-negative (E41I) RHO1 alleles. The dominant active mutation was inserted with Rho1_GV1 and Rho1_GV2 to obtain pNV40 and the dominant negative mutation was inserted with Rho1_DN1 and Rho1_DN2 to obtain pNV41. To generate BNI1 expression constructs, genomic sequences of *bni-1* were amplified from bacmid BAC NC20 K10 (Fungal Genetic Stock Center) and were inserted into pNV47, a modified pNV80 that encodes a N-terminal MYC^3 fusion. For *bni-1*$^{aa1\text{-}824}$ encoding the dominant negative BNI1 N-terminus the primers NV_SepA_1a and NV_SepA_7, for *bni-1*$^{aa1029\text{-}1817}$ encoding the dominant active BNI1 C-terminus the primers NV_SepA_6 and NV_SepA_5 were used, resulting in expression plasmids pNV83 (containing *bni-1*$^{1\text{-}824}$) and pNV84 (containing *bni-1*$^{1029\text{-}1817}$).

2.3 General molecular methods

2.3.1 Bioinformatics

Alignments were performed with MultAlin (Corpet, 1988; http://bioinfo.genopole-toulouse.prd.fr/multalin/multalin.html). Structure predictions were used from Cn3D of the NCBI to study the 3D structure of protein domains. This information helped to design truncated proteins for expression and ensure the use of a complete structural conserved domain in biochemical studies.

2.3.2 General cloning procedures

Standard methods were performed as described (Ausübel *et al.*, 1997; Sambrook *et al.*, 1989) or according to manufacturers instructions with minor modifications. For analytical PCR 10 fold concentrated buffer (200 mM Tris, pH 8.8, 100 mM KCl, 100 mM $(NH_4)_2SO_4$, 22.5 mM $MgCl_2$, 0.02% Nonidet P40 (NP40) and 0.02% Triton X-100, 40% glycerol) was used. The buffer composition was adjusted to the melting behaviour of the DNA by varying the di-methyl-sulfoxide (DMSO) content up to 10% of the PCR volume for GC-rich templates. For most PCR reactions, initial denaturation was performed for 2 min at 94°C, followed by 30 cycles. A cycle consists of 30 s denaturation at 94°C, annealing for 30 s, extension for time dependent on the polymerase (1 min/kb for Taq-polymerase, 2 min/kb for Pfu-polymerase, for other polymerases according to manufacturers instructions) and fragment length dependent time at 72°C. For long fragments, a terminal extension of 10 min was added in preparative PCR reactions. Mutagenesis PCR from the pNV2 plasmid with a size of 9 kb was done with Pfu-polymerase in 18 cycles including 20 min extension time each. For analysis of transformants of *E. coli* or yeast, colony PCR was done. *E. coli* colonies were transferred with a sterile toothpick into the 20 μl PCR reaction. Yeast colonies were suspended in 15 μl 10 mM sodium hydroxide and boiled for 10 min at 94°C. Afterwards yeast fragments including genomic DNA were pelleted by centrifugation at 13000 rpm for 2 min. 1 μl of the supernatant was used as template in 50 μl PCR reactions. For yeast colony PCR 35 cycles were used.

Ligation reactions were performed either in restriction buffers Yellow or Red (Fermentas, Vilnius, LT), supplemented with 1 mM ATP for at least 1 h to over night, or with 2 fold concentrated quick ligation buffer (100 mM HEPES, pH 7.6, 20 mM $MgCl_2$, 4 mM DTT, 4 mM ATP and 14% (v/v) PEG4000) for 3 to 30 min.

For subcloning of PCR fragments, TOPO-TA and TOPO zero blunt cloning kits (Invitrogen, Carlsbad, USA) or Gene Jet and Clone Jet cloning kits (Fermentas, Vilnius, LT) were used. pNV86 (see supplementary data) was occasionally used for subcloning of PCR fragments to reduce the background of further cloning steps.

2.3.3 Immunological methods

For the preparation of polyclonal anti-POD6 and anti-LRG1 antibodies, a POD6 fragment consisting of amino acids 421–675 as hexa-histidine (His6)-fusion protein (encoded from pNV74) and a LRG1 fragment consisting of amino acids 781 – 848 as maltose binding fusion protein (encoded from pNV25) were used as antigens, respectively.
Pineda Antikörper Service (Berlin, Germany) generated polyclonal antisera. The sera were affinity-purified using the original antigen of anti-POD6 sera or a gluthatione S transferase (GST) fusion of the LRG1 amino acids 781 – 848 encoded from plasmid pNV26 for anti-LRG1 sera, respectively.
The expression of the MYC-tagged LRG1 constructs was verified by Western Blot analysis using monoclonal 9E10-anti-cMYC antibodies (Santa Cruz Biotechnology, Santa Cruz, USA).
For protein extraction, *N. crassa* mycelia samples were frozen and grinded in liquid nitrogen using mortar and pestle. The powder was suspended in 50 mM phosphate buffer, pH 7.0, 150 mM KCl, 1 mM DTT, containing protein inhibitors (0.5 mM phenylmethylsulphonyl fluoride (PMSF), 350 μg/ml benzamidin, 10 μg/ml aprotenin, 10 μg/ml leupeptin and 2 μg/ml pepstatin A). Phosphatase inhibitors (1 mM sodium fluoride, 25 mM β-glycerophosphate and sodium orthovanadate that was prior to use incubated for 5 min at 94°C at a final concentration of 10 mM) were added to the buffer for Western Blot experiments probed with anti-phospho specific antibodies (Cell Signaling Technology, Danvers, USA). Centrifugation of the samples was done at 13000 rounds per minute for 30 min at 4°C. The protein containing supernatant was used for further experiments. Proteins were separated by SDS-PAGE using 7.5%, 10%, 12% or 15% gels. Western blotting was performed according to standard procedures (Ausübel *et al.*, 1997). The phospho-p44/42 MAP kinase (Thr202/Tyr204) antibodies (Cell Signaling Technology, Danvers, USA) were used according to the manual.

2.4 Biochemical methods

2.4.1 Protein purification

The POD6 fragment consisting of amino acids 421–675 was expressed from plasmid pNV74 in *Escherichia coli* BL21 (DE3) as a pQE60-based His6-fusion protein and purified via the Ni-NTA purification system (Qiagen, Hilden, Germany). For the preparation of polyclonal anti-LRG1 antibodies, a LRG1 fragment containing amino acids 781 – 848 was expressed from plasmid pNV25 in *Escherichia coli* BL21 (DE3) as a MalE-fusion protein and was purified via affinity purification using amylose agarose (NEB, Ipswich, USA).

The RHO GTPases and CDC24$^{204-544}$ were purified as GST fusion proteins from *Escherichia coli* BL21 (Rosetta 2). Cells were inoculated to an OD_{595} of 0.1, grown at 20°C to an OD_{595} of 0,45 and induced for 4 h with 0.1 mM IPTG. Cell extracts were generated by sonification in lysing buffer (50 mM Tris, pH 7.5, 10% sucrose, 5 mM $MgCl_2$, 1 mM PMSF, 0.008% ß-Mercaptoethanol, 0.02% NP40). Proteins were bound to GSH sepharose (Amersham, Buckinghamshire, United Kingdom), washed in washing buffer (50 mM Tris, pH 7.5, 250 mM NaCl, 5 mM $MgCl_2$, 1 mM PMSF, 0,008 % ß-Mercaptoethanol, 0.02% NP40) and eluted with elution buffer (50 mM Tris, pH 8.0, 250 mM NaCl, 5 mM $MgCl_2$, 5 mM DTT, 20 mM glutathione (red.), 0,02% NP40).

2.4.2 Enzymatic assays

2.4.2.1 *In vitro* assay for Rho GAP activity

GTPase assays were performed as described (Gibbs, 1995; Gibbs *et al.*, 1988; Morii *et al.*, 1991). Pre-equilibration of the RHO proteins was performed for 15 minutes at 25°C in 30 μl buffer A (20 mM HEPES, pH 7.6, 25 mM sodium chloride, 2 mM EDTA, 1 mg/ml BSA, 0.5 mM DTT, 0.005% cholic acid) including 0.5 μM GTPase and 5 μCi (0.17 μM) [γ-^{32}P]-GTP. Adding 1 μl 0.5 M $MgCl_2$ and putting the samples on ice stopped GTP loading. 5 μl GTPase was added to start the reaction (final concentration: 20 mM HEPES, pH 7.6, 1 mg/ml BSA, 0.1 mM DTT, 1 mM GTP and 4 μM GST::GAP or purified GST) at 25°C. 5 μl samples of the reaction were stopped in 1 ml ice-cold wash buffer (50 mM Tris, pH 7.5, 50 mM sodium chloride, 5 mM $MgCl_2$) and filtered trough BA85 nitrocellulose membranes (Whatman, Maidstone, Great Britain). The filters were washed with 6 ml wash buffer, dried and measured by scintillation counting in a QuantaSmart scintillation counter (Perkin Elmer, Waltham, USA).

2.4.2.2 *In vitro* assay for Rho GEF activity

Initially in cooperation with PD Dr. Jan Faix, Hannover Medical School, fluorescence of a GDP nucleotide exchange reaction using the modified fluorescent nucleotide 2`, 3`-O-(N-methylanthraniloyl)-Guanosin (Mant GDP) was measured in a Jasco FP–6500 spectrofluorometer. Further experiments were performed in a TECAN plate reader (Tecan Trading AG, Switzerland). Excitation at 356 nm was used and the emission at 448 nm was measured for 600 seconds. The samples contained 1 μM GST::RHO protein (whereas RHO means *Neurospora* RHO1 to RHO4,

RAC or CDC42; purified proteins) 0.1 μM Mant GDP, 30 mM Tris–HCl, 5 mM $MgCl_2$, 10 mM KH_2PO_4, 3 mM DTT, pH 7.5 at 21°C and as control 1 μM GST or for stimulation of the activity 1 μM of GST::CDC24$^{204-544}$.

2.5 Microscopy

Low magnification documentation of fungal hyphae or colonies was performed with a SZX12 stereomicroscope (Olympus, Tokyo, Japan) and a PS30 video camera (Kappa opto-electronics GmbH, Gleichen, Germany).

2.5.1 Immunofluorescence

Immunolocalization for *N. crassa* hyphae was conducted following a protocol adapted from Minke *et al.* (1999). Conidia were grown for 12 h on a small piece of GN-6 cellulose filter (Pall Corporation, New York, USA) placed on the surface of an agar plate. Filters were plunge-frozen in liquid propane and transferred to fixative (4% formaldehyde in 96% ethanol pre cooled to –80°C). Samples were maintained at –80°C for at least 2 days and then slowly transferred to room temperature (2 h at –20°C, 2 h at 4°C, room temperature). Filters were rehydrated in a series of ethanol:buffer (100 mM phosphate, pH 7.0) solutions starting at a ratio of 90:10 and ending at 10:90. After several rinses in phosphate buffer, the cell wall was digested by incubating the filters for 5 to 8 min in 2 mg/ml lysing enzymes from *Trichoderma harzianum* (Sigma, St. Louis, USA) in buffer containing 100 mM potassium citrate (pH 6.0), 20 mM EGTA and 5% BSA. The lysing enzymes were washed away by several short rinses with phosphate buffer. To block non-specific binding of antibody, the filters were incubated for 1 h in 5% BSA in PBS. Samples were immersed in the primary antibody in 5% BSA/PBS for at least 8 h, washed several times in phosphate buffer, incubated with the secondary antibody for ≥ 8 h and washed again for at least 4 h with several changes of the phosphate buffer. Samples were visualized using standard DAPI, rhodamine and FITC filter sets. For immunolocalization, samples were viewed with an ORCA ER digital camera (Hamamatsu Photonics, Hamamatsu, Japan) mounted on an Axiovert S100 microscope (Carl Zeiss AG, Oberkochen, Germany). Image acquisition was done using the Openlab 5.01 software (Improvision, Coventry, Great Britain) and images were further processed using Photoshop CS2 (Adobe, San Jose, USA).

2.5.2 GFP fluorescence

To visualise GFP fluorescence of growing hyphae, strains were grown on a 2 to 5 mm layer of

Vogels medium containing 2% agar and 0.1 μg/ml panthothenic acid on top of a second 1 cm agar layer at 25°C. Sections of about 5 mm x 8 mm were cut at the edge of colonies and incubated in water for about 10 min. Coverslips that were stuck to holes in small plates were used as slides. The agar blocks were put on these slides with the colonies directed towards the slip and 40 μl of water or for chitin staining 40 μl of 2 μg/ml Calcofluor White were added.
For localization studies of GFP::LRG1 and GFP::LRG1*, an AxioObserverZ.1 microscope equipped with an ApoTome unit, an AxioCam MRm r3.0 CCD camera and X-Cite (EXFO) illumination source with filter sets 1 and 38 was used with AxioVision Software 4.6. For confocal imaging an Axiovert 100M microscope with the confocal module LSM 510 and LSM 510-Software was used (all from Zeiss, Germany). Live time measurements were performed using an ORCA ER digital camera (Hamamatsu Photonics, Hamamatsu, Japan) mounted on an Axiovert S100 microscope (Carl Zeiss AG, Oberkochen, Germany). Image acquisition was done using the Openlab 5.01 software (Improvision, Coventry, Great Britain) and images were further processed using Photoshop CS2 (Adobe, San Jose, USA).

3 Results

At the starting point of this work, COT1 was the best-characterized regulator of hyphal elongation in *N. crassa*. Mutants in genes, which were specifically required for polar tip extension (Seiler and Plamann, 2003) were analysed to obtain evidence for their functional relationship. *pod-6* and *lrg-1* mutant alleles were chosen for a phenotypic characterization and a *cot-1* mutant strain with similar growth defects was selected for a comparative analysis.

3.1 POD6 and LRG1 are essential for hyphal tip extension

RT-PCR and sequencing experiments of cDNA with POD6-specific primers were used to define *pod-6* as a 928-amino acid protein containing an N-terminal kinase domain with the highest sequence similarity to members of the Ste20 group of kinases and a C-terminal region with no characteristic sequence motifs (Seiler *et al.*, 2006). Based on the homology of the kinase domain, its N-terminal localization and the lack of defined sequence motifs in the C-terminus, POD6 belongs to the GCK-III subfamily of eukaryotic Ste20 kinases. The known vertebrate members of this subfamily, SOK1, MST3, and MST4, have been implicated in the regulation of stress response, apoptosis and proliferation, but the molecular mechanisms are unknown (Dan *et al.*, 2002; Dan *et al.*, 2001; Lin *et al.*, 2001; Qian *et al.*, 2001). The most closely related budding and fission yeast kinases are Kic1p and Nak1, respectively. Both kinases have been reported as part of a morphogenetic network that also contains the NDR kinases Cbk1p (budding yeast) and Orb6 (fission yeast), respectively, that is important for coordinating polarized growth with daughter cell specific transcription and cell cycle progression (Kanai *et al.*, 2005; Leonhard and Nurse, 2005; Nelson *et al.*, 2003).

When wild type *N. crassa* conidia (asexual spores) germinate, they rehydrate and begin to grow isotropically. Growth soon becomes polarized, and usually one hyphal tip is generated. Continued polarized growth results in unidirectional extension of the straight primary hypha and branching from subapical compartments subsequently generates new hyphal tips. Continuous hyphal elongation and branching results in the formation of spreading colonies (Collinge and Trinci, 1974; Momany, 2002; Seiler and Plamann, 2003).

To determine the morphological changes conferred by the conditional mutations in *pod-6* and *lrg-1*, a microscopic analysis of different temperature sensitive alleles of *pod-6* and *lrg–1* grown under permissive and restrictive temperatures was conducted. As no major morphological differences were observed among the seven isolated *pod-6* alleles and the 22 *lrg-1* alleles of the screen (Seiler and Plamann, 2003), *pod-6(31-21)* and *lrg-1(12-20)* alleles were chosen for further investigations.

While no differences in hyphal elongation or branching frequency were detected between *pod-6(31-21)* and *wild type*, differences could be measured between *lrg-1(12-20)* and *wild type* grown at permissive temperatures (19°C to 25°C) upon careful examination (Figure 3). An increased number of branches were detected within the apical 500 μm of *lrg-1(12-20)* resulting in denser colonies at 19°C and 25°C. In addition, the elongation rate of *lrg-1(12-20)* was slightly reduced compared to *wild type* at 19°C (0.74 ± 0.16 μm/h and 1.7 ± 0.3 μm/h, respectively). This difference was more prominent at 25°C (0.87 ± 0.2 μm/h and 3.45 ± 0.26 μm/h, respectively). The diameter of hyphae that undergo subapical branching at the edge of a colony was reduced to 6.2 ± 1.0 μm (n=65) for *lrg-1(12-20)* compared with 9.8 ± 2.1 μm for *wild type* (n=10) at 25°C. Thus, the function of the mutant LRG1 is already partially impaired at permissive temperature in *lrg-1(12-20)*.

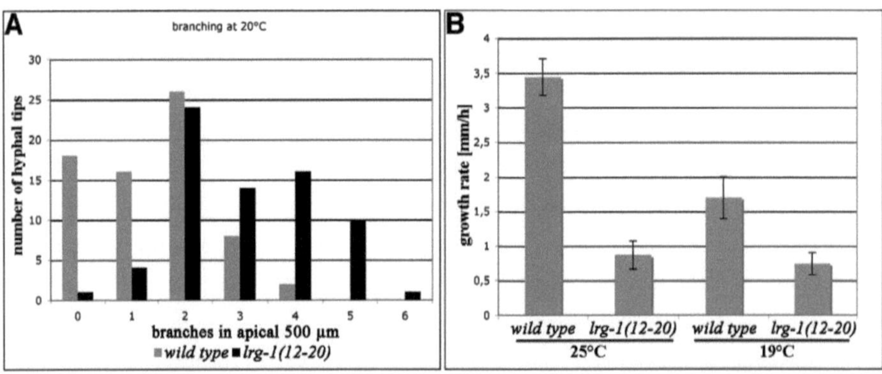

Figure 3: Morphological characteristics of *lrg-1(12-20)* and *wild type* at permissive temperature.
A) The number of branches within the apical 500 μm of 70 hyphal tips were analysed. On average 1.4 branches for *wild type* compared to 3.1 branches for *lrg-1(12-20)* were detected. B) The radial growth rates of the indicated strains were measured at permissive temperatures (n=5).

A rapid cessation of tip extension was observed within 30 min after transferring *lrg-1(12-20)* and *pod-6(31-21)* to 37°C, while the temperature shift had no effect on the morphology of *wild type* (Figure 4). This cessation of tip extension of *lrg-1(12-20)* and *pod-6(31-21)* was accompanied by the appearance of numerous subapical needle-shaped branches of ≤10 μm in length in case of *lrg-1(12-20)* and of 20-50 μm in length for *pod-6(31-21)*, which also stopped growth with a pointed tip. After prolonged incubation at restrictive temperature, the new branches as well as the primary hyphae began to swell up in a bulbous and apolar manner in both strains. Transfer of such a culture back to permissive temperature resulted in tips resuming normal growth rates, diameter and morphology within 30 min for *lrg-1(12-20)* and within a few minutes for *pod-6(31-21)*. These experiments indicate that LRG1 and POD6 are both essential for tip extension, but are also required to inhibit excessive branch formation in subapical regions of the hyphae.

The *cot-1(1)* mutation has been shown to result in increased cell-wall thickness at restrictive temperature (Collinge *et al*., 1978; Gorovits *et al*., 2000), suggesting a defect in cell-wall metabolism. Chitin is the primary component of fungal septae and of the inner layer of the hyphal cell wall, and is therefore accessible to the Calcofluor White staining primarily at the hyphal tips and at septae. To investigate chitin deposition in the *lrg-1(12-20)* and *pod-6(31-21)* mutants Calcofluor White staining was used (Figure 4). At permissive temperature, hyphal tips (and to a minor extent also septa) of all three mutants were strongly labelled as in the *wild type*. After a shift to 37°C of the *lrg-1(12-20), pod-6(31-21)* and *cot-1(1)* mutants grown at permissive temperature, for 1 h in case of *lrg-1(12-20)* and for 5 h in case of *pod-6(31-21)* and *cot-1(1)*, the mutants showed extensive labelling. This staining occurred in a patchy, subapical fashion throughout the hyphae at positions of newly emerging branches, including strong septal staining, and indicates excessive chitin deposition at sites of aberrant growth in all mutants. These similar cell wall alterations further suggest a functional connection between POD6 and COT1. Furthermore, the strong calcofluor staining at the tip observed in mutants grown at the permissive temperature is nearly absent when the strains are cultured at the restrictive temperature for the mentioned times, indicating decreased chitin synthesis at these tips, which are no longer elongating. Septation increases after shift of *lrg-1(12-20)* and *pod-6(31-21)* to restrictive temperature. The length of hyphal compartments of *lrg-1(12-20)* were 53.7 μm +/- 21.0 μm (n = 50; for *wild type* 148 +/- 58 μm) when the strain was grown at 25°C, while after the transfer to 37°C for 2.5 h the compartment length was reduced to 11,3 μm +/- 4,3 μm (n = 50).

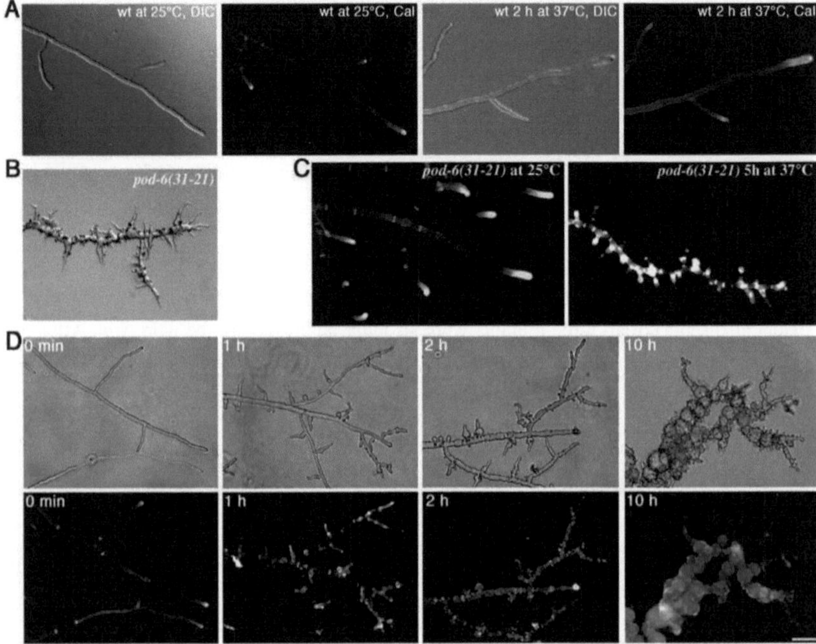

Figure 4: LRG1 and POD6 are necessary for coordination of tip growth, branching and septation in *N. crassa*.

(A) Wild type *N. crassa* was grown at 25°C and shifted to 37°C for two hours. DIC images show the morphology of the hyphae. Calcofluor White (Cal) stained the apex of the tips and septae. (B) A colony of *pod-6(31-21)* was grown on agar at room temperature (25°C) over night and shifted for 10 h to 37°C as the restrictive temperature. The hyphae show a typical barb-wired phenotype with numerous subapical branches and tapered tips. (C) Calcofluor White stained primarily the hyphal apex of *pod-6(31-21)* grown at permissive temperature but after shift to restrictive temperature for 5 h the strain showed an abnormal and patchy labelled staining, including strong septal staining throughout the hyphae. The tip staining disappears at restrictive temperature. Pictures adapted from (Seiler *et al.*, 2006). (D) *lrg-1(12-20)* was grown on minimal media plates at 25°C and shifted to restrictive temperature for the indicated times to illustrate the cessation of tip extension with pointed, needle-like tips and the progressive hyperbranching of the mutant (phase contrast images, upper panel). Increased septation and abnormal chitin distribution was monitored by staining with Calcofluor White (lower panel). Scale bar for all pictures is 40 μm.

3.2 *pod-6* and *lrg-1* deletion mutants show identical phenotypes like conditional mutants

To examine the phenotype of *pod-6* and *lrg-1* deletion strains, mutants were constructed by using the „sheltered disruption“ method (Nargang *et al.*, 1995), which takes advantage of the fact that *N. crassa* is a multinucleated cell (Springer and Yanofsky, 1989). A schematic presentation of the generation of the deletion cassette is shown in Figure 5.

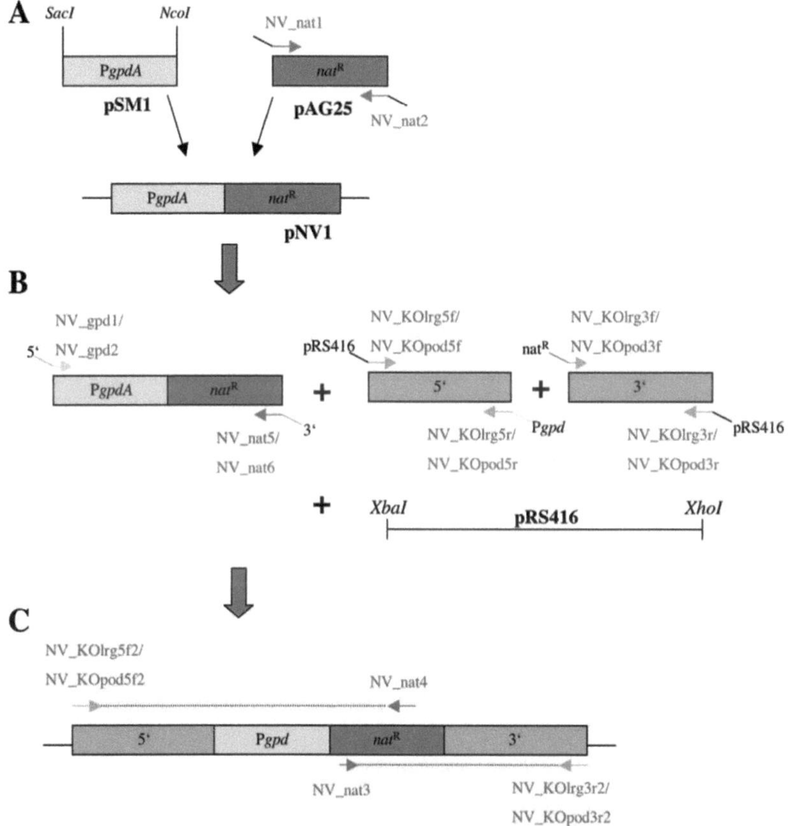

Figure 5: Generation of the deletion cassettes for *lrg-1* and *pod-6*.
(A) The new resistance marker plasmid pNV1 was established for *N. crassa*. The *A. nidulans gpd*A promoter derived from pSM1 was inserted into pBluescript SK+ together with a nourseothricin resistance (*nat*^{R}) gene amplicon derived from pAG25. (B) The resistance marker cassette and the flanking regions of *lrg-1* or *pod-6*, respectively, were amplified and transformed into *S. cerevisiae* together with the linearized yeast vector pRS416 to obtain the deletion cassettes. (C) The deletion cassettes were transformed as two PCR amplicons that do not mediate the nourseothricin resistance on their own. Primers are indicated in red, homologous regions as primer extensions and restriction sites in black letters.

The *pod-6* and *lrg-1* genes were deleted via homologous recombination, which was forced by transformation of a split marker cassette fused with the 5' and 3' regions of the respective genes. The newly generated nourseothricin resistance marker cassette (*nat*^{R}) mediates an antibiotic resistance up to 150 μg/ml for several independent ectopic integrations, compared to a maximal tolerance of 7.5 μg/ml for the untransformed *wild type*. The resistance cassette was recombined in yeast with flanking regions of *lrg-1* and *pod-6* and then transformed as two overlapping PCR fragments, which were only able to mediate a resistance after recombination, in the *N. crassa* heterokaryotic strain HP1 (Nargang *et al.*, 1995). As a result of the spilt marker cassette, the number of resistant transformants was reduced approximately 100 fold when compared to transformations of the same amount of plasmid DNA. 20% and 40% of the occurring transformants of *Δpod-6* and *Δlrg-1*, respectively, showed homologous integration events as proven by PCR and complementation analysis (data not shown). The use of the heterokaryotic strain HP1 allowed the generation of mutants in genes that are essential or important for growth. The resulting mutant harboured two kinds of nuclei, one with a null allele of *pod-6* or *lrg-1*, respectively, and one with a *wild type* copy. These nuclei contain selectable markers that allowed shifting the nuclear ratio in the heterokaryotic cells. All obtained deletion strains contained the null allele in the benomyl resistant nucleus (71-18). Growth on media containing p-Fluoro-DL-phenylalanine (fpa) and histidine favoured the propagation of the *wild type* nucleus. In contrast, growth of heterokaryotic cells on media containing benomyl and pantothenic acid forced the knockout nucleus of *Δpod-6* or *Δlrg-1* to predominate, and thus resulted in the depletion of POD6 or LRG1, respectively (for details see section 2).

The morphological defects observed under these conditions were indistinguishable from *pod-6(31-21)* or *lrg-1(12-20)*, respectively, when these strains were germinated at restrictive temperature (Figure 6). These results indicate that *pod-6(31-21)* and *lrg-1(12-20)* are temperature-sensitive loss-of-function alleles of *pod-6* and *lrg-1*, respectively. The phenotypes of the two mutants further indicate an essential role for POD6 and LRG1 during the extension of the hyphal tip and in controlling the number and position of subapical branches similar to what has been reported for COT1. In addition to these significant phenotypic similarities between the three mutants, there are morphological differences that distinguish *lrg-1* from *pod-6* and *cot-1*. The tips of *lrg-1* mutant strains are much thinner and shorter as those of the *cot-1* and *pod-6* mutants, which suggests distinct functions of LRG1 and COT1/POD6.

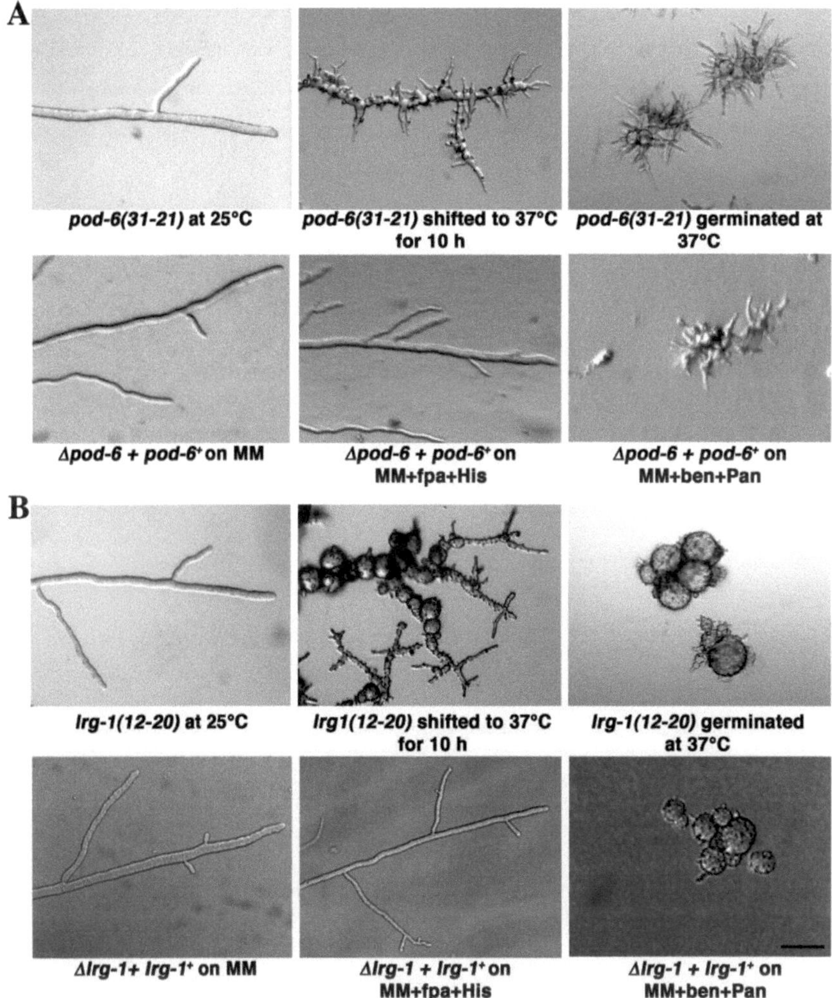

Figure 6: POD6 and LRG1 are essential for hyphal tip extension.

(A) Comparison of *pod-6(31-21)* morphology (upper panel) with *Δpod-6 + pod-6⁺* (lower panal). (B) Comparison of *lrg-1(12-20)* morphology (upper panel) with *Δlrg-1 + lrg-1⁺* (lower panal).

Conditional mutants were grown on minimal medium plates at 25°C and shifted to restrictive temperature for 10 h to illustrate the cessation of tip extension with pointed, needle-like tips and the progressive hyperbranching of the mutants. When germinated at restrictive temperature, the mutants exhibited a compact, hyperbranched colonial morphology. The growth of heterokaryotic deletion strains on the indicated media resulted in wild type or deletion phenotypes. In both cases the gene replacements occurred in the nucleus containing the benomyl (ben) and pantothenic acid (Pan) markers, and growth under conditions that select for this nucleus (presence of ben and Pan) resulted in morphological defects identical to those of the conditional mutants germinated at restrictive temperature for 15 h. Scale bar is 7.5 μm for all pictures.

To facilitate localisation studies and biochemical approaches an antiserum against a $POD6^{421-675}$::His6 fusion construct was generated and affinity-purified. The antibody recognized a single polypeptide of ca. 100-kDa in *wild type* extracts (Figure 7 A). This signal was strongly reduced in extracts of the heterokaryotic *Δpod-6* strain grown on benomyl and pantothenic acid, indicating that the generated anti-POD6 antibody is specific for POD6 (Figure 7 B).

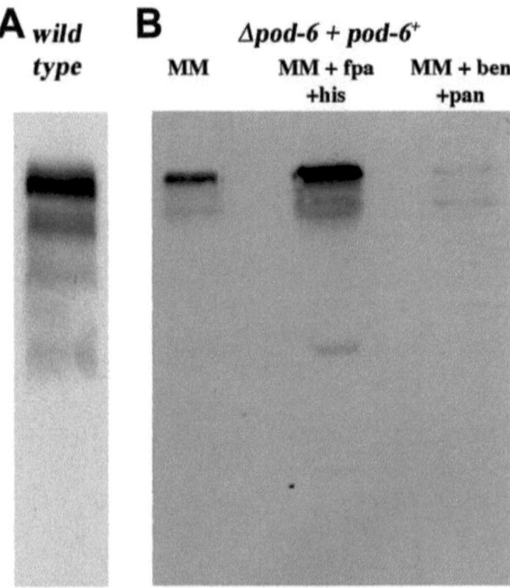

Figure 7: Affinity purified antibody specifically detects POD6.
(A) POD6 is detected in *N. crassa* extract prepared from *wild-type* cells. (B) Cell extracts of the nuclear-ratio-modulated heterokaryotic *Δpod-6* + *pod-6*$^{+}$ strain grown in the indicated media were adjusted to equal amounts of soluble protein and probed with anti-POD6 antibody to determine the effect of the predominance of *Δpod-6* nucleus in the presence of benomyl and pantothenic acid on POD6 abundance. The blot illustrates the specificity of the used antibody.

3.3 COT1 and POD6 act together to regulate polar tip growth

Germination of *lrg-1(12-20)* and *pod-6(31-21)* at restrictive temperature resulted in the formation of compact colonies with multiple ≤10 or 10-50 µm long germ tubes with pointed tips, respectively. Furthermore, *pod-6(31-21)* produced secondary and tertiary branches (Figure 8 A).

The defects of *pod-6(31-21)* were strikingly similar to the known morphological defects of *cot-1(1)* (Collinge *et al.*, 1978; Collinge and Trinci, 1974; Terenzi and Reissig, 1967; Yarden *et al.*, 1992), indicating that POD6 and COT1 may have related functions.

To further analyze the functional relationship between COT1 and POD6, a *cot-1(1);pod-6(31-21)* double mutant was generated (Figure 8 B). The *cot-1(1)*, *pod-6(31-21)* and *cot-1(1);pod-6(31-21)* mutants showed identical phenotypes on minimal media agar plates at restrictive temperature. This provides further evidence of the involvement of *cot-1* and *pod-6* in a common genetic pathway. In contrast, *cot-1(1);lrg-1(12-20)* double mutants were synthetically lethal, suggesting the involvement in different pathways with a common function.

Figure 8: Growth phenotypes of temperature sensitive mutants of *cot*-1, *pod*-6 and *lrg*-1 at restrictive temperatures.
(A) Conidia of the indicated strains germinated on solid medium for 12 hours at 37°C. Scale bar is 7.5 μm. (B) A *cot-1(1);pod-6(31-21)* double mutant displayed the same morphological defects as the two parental strains. All strains were grown on minimal media plates shifted to restrictive temperature for 5 h after growth at permissive temperature for 10 h. Pictures in (B) kindly provided by Stephan Seiler.

An increased thickness of the cell wall and the accumulation of small vesicles was previously reported in ultra structural analyses of *cot-1(1)* transferred to restrictive conditions (Collinge *et al.*, 1978; Gorovits *et al.*, 2000). Therefore, it was tested whether the observed phenotypic similarities between *cot-1(1)* and *pod-6(31-21)* extend to the ultra structural level. Transmission electron microscopy was used to determine the ultra structure of *wild type* and *pod-6(31-21)* shifted to 37°C for 10 h. A strongly thickened cell wall and the presence of abundant small vesicles were observed (Figure 9).

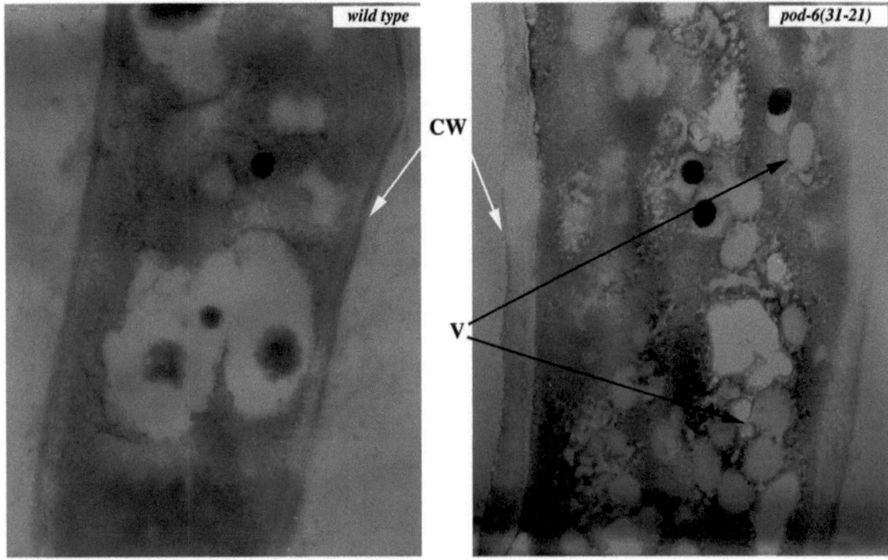

Figure 9: The cell wall of POD6 is thickened.
Electron microscopic images of the indicated strains grown at 25°C for 8 h and shifted to 37°C for 10 h reveal an increased thickness of the cell wall (CW) and the presence of abundant vesicles (V) in *pod-6(31-21)*.

The identical morphological defects, the common increase of cell wall thickness in electron microscopic images and the morphology and chitin distribution of the double mutant at restrictive temperature indicate the involvement of *pod-6* and *cot-1* in a common pathway. This might be a hint for a physical interaction of the proteins.

A potential physical interaction between COT1 and POD6 was examined in a strain where a *myc::cot-1* transgene was ectopically integrated (kindly provided by Carmit Ziv, The Hebrew University of Jerusalem, Rehovot 76100, Israel). S. Seiler performed immune precipitation experiments and localization studies using the purified α-POD6 antibody. These studies showed a physically association and a partial colocalization of POD6 and COT1. The localization was dependent on opposing microtubule-dependent motor proteins (Seiler *et al.*, 2006).

3.4 The function of LRG1 in polar tip elongation

3.4.1 LRG1 is a fungal specific protein containing LIM and GAP domains

Examination of the 1279 amino-acid (aa) LRG1 encoding sequence revealed two interesting features. The N-terminal region from aa 1 – 533 is cysteine- and histidine-rich and contains three LIM domains. These domains are tandem zinc-finger containing structures that act as versatile protein-protein interaction motifs (Bach, 2000; Brown and Turner, 2004; Kadrmas and Beckerle, 2004; Schaller, 2001). Apart from the fungal LRG1 relatives, the LIM domains of LRG1 are most closely related to the focal adhesion organizing protein paxillin with an E-value of $2e^{-28}$ for *Drosophila pseudoobscura* paxillin. Furthermore, a RHO-GAP domain was found in the C-terminal part from aa 791-1279 of LRG1. An alignment of fungal LRG1 homologues is shown in Figure 10. Due to a lack of experimental data, *in silico* predictions were used for the annotated sequences. The RHO-GAP domain (red box) as well as the LIM domains (green boxes) are highly conserved, while the amino acids between these parts are more divers, indicating that the conserved domains of LRG1 may have important functions throughout the fungal kingdom. In case of the basidiomycete *Coprinopsis cinerea okayama* the first LIM domain is missing.

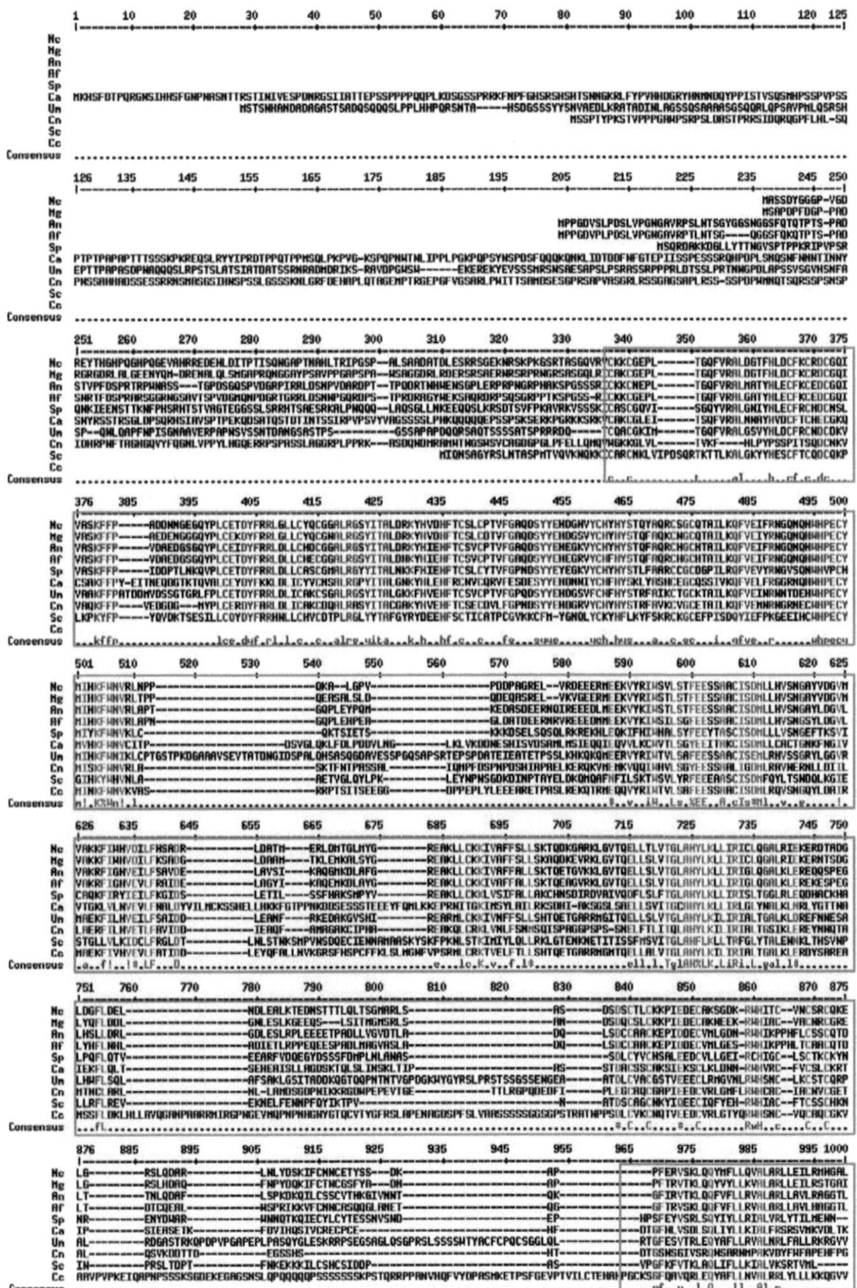

```
           1001 1010      1020      1030      1040      1050      1060      1070      1080      1090      1100      1110      1120 1125
           |--------+---------+---------+---------+---------+---------+---------+---------+---------+---------+---------+---------+----|
       Nc  PSAAEARGHNGIDAGDGRPLMDGRQRGFPGGDQ-------------------SLHHRE----SSYESAL-------------------------------------NDVKRLKSTRLDKHLSSS
       Mg  PREDGHPQDLHDGQRDLPPHLRSEARSKSFAAG--------------------SMHQRE----SSYESAV-------------------------------------NDVKRLKSTRLDKALSSS
       An  FPPETD--MPSSFVEDSVVQSDGNQIPPGGELR-------------------RRKDGDG---SSLEQTV-------------------------------------GEMRRLRSIRNERTLSTT
       Af  QPTSDDPSFAGSETQDGNQTQSGGNIHRSTTRS-------------------KYAGRDGAAESSLEQTV-------------------------------------GEMRRLRSIRNERTLSTT
       Sp  DSSQRKPLSVDPKQENVSSTVETAKQAETTASS-------------------DSFRK-------YANTL-------------------------------------NDLRHLKSSRNRKATSNE
       Ca  VPRSYLSDSQS--DTDTVDVAQDNYSQTLNDVTSLRSKRQSQ--------KLSKSIKKKA---RKSIIVEAPEADKARKEDIKTHP---LGMKS--------SGVGTNDTSRSSGINDETTIADE
       Um  PPSPPVSATRQVGGTDTSTPGAASPADANEQVSMHEAYRDSQDIKRMKSVNLNRKLSTKAKVPRISTVVGSPSGRQTQTSDMQKHSPSDIGLQTDSRRSPRQASPGSSPREPSSPFKREVGSSRE
       Cn  MVKPPEKIWCGIHKNPNGSWKGFQGVSRLEQYAFL----------------LHIALRRLYVHFRLHHSLQS----------VRDH-----GM----------SGRAESEVKRMKSVTLDRKLSST
       Sc  --KSKASNKVGRNSLQSTMLKEHTYIRTL-------------------------------------------------------------------------NDIKRLRSRRESVRVTHN
       Cc  PASPATTTFDQQQVQPPGQQGGQQAGQGRDLYTYYLDRKLSATAR-----LPKRSTVVESPTGKTPDSAESQRYYQDGTPGQQAPSSAAPTSSGYSSPGPSSTSYPSTPYRPPLPSQQQAAQVQQ
Consensus  .............................................................................................................................

           1126 1135      1145      1155      1165      1175      1185      1195      1205      1215      1225      1235      1245 1250
           |--------+---------+---------+---------+---------+---------+---------+---------+---------+---------+---------+---------+----|
       Nc  FRKARTSRIMDGPESSSAQPGSGPGSGIDGGPANGRRMRI--------EE--DMGPGDDHDMMLPNQDALTLADIPRIVAAEQAREQRTGPPSRHELLRSPATEPSFGGASSGSGHKRSYSDSKG
       Mg  VRKARTSRIMEGPDGTAGRPGS---AGANSGDPRKPGFSI--------VE--DPANERAEDSMLGNQDALTLDDIPRIVAAEQAREQRPNAFQRQELFRAPATEPNLN-----SNNPRAFSPGRG
       An  YKKARASRIIDGPEGRSVRPGS---SGGEGSDSRGPGFQI--------VEEKDANGEPVTELAFGNQDALTLDDIPRIVAAEQAKEQRPNAYRHAGTKLVGTTEPLPR---YNQGHQRGVSNAGI
       Af  YKRARASRIIDGPEGRSVRPGS---SGGEGSDARGHGFQI--------VEERDANGETVTDLTFGNQDALTLDDIPRIVAAEQAKEQRPNAYRHAGTKLVGTTEPLPR---YNQGHQRGVSSGNL
       Sp  TQSSSNSTETSKLSKNVSESGKDKSPHWHSHGGSITGKSI--------VEQASSPLERRMDAFDENR-AFTLDDIPKVISEQRNREHRPNAFRHMPSYTDPSYRKNSG-----AIYDKNDGTQKG
       Ca  LNDLSFDSQTSKQQGSRKTSSASQLSYNPGGDENLSVTRKSLKIRDEPQRQTTNTHLDRTSDLLKNEKSLTLDDIPRIVAAEQAREQRPNAFKHHNSLYQRQTAPHRL---KATGHTSTVITPTG
       Um  PSPSSRARQAQPPQGYQQQQQQQQQPYKQQQVFQPVQQQDHLPFALPPQQQQPPQFGSGTSGRSGSPSNL---EPGSIVPIRPAFARHNTDVKIREDGPMRQPSGDEI---QRETRSEDGITLAD
       Cn  ARLPQRSMVVESPAGRMADGNGQIISARNAESPTPTPTPK------EPQTEIIITTEDASNNAT------------TVNVLRPPFARNNTSVKIIKENSGESPQRGFR---RVSAPKEDDAITLG
       Sc  KQQARKSVILETAETDLNDPTK-------QGDSKNLVIQT--------DDPSSSQQVSTRENVFSNTKTLTLDDISRIVAAEQARELRPNAFAHFKKLKETDDE------------TSNVVPKKS
       Cc  FQQQQQQQQQQQMTPRAHPGKLLDPRGAGMDQGQRSGYDSMPSSPAGGPGDYSHSHDSYQSYPHTQSSTSSYHSQSQTPSSHPQYSPYPHSQHSQYPSPPSHTSYQHSNHQRHQKSLDDGITLAD
Consensus  ...........................................................................................P.................................

           1251 1260      1270      1280      1290      1300      1310      1320      1330      1340      1350      1360      1370 1375
           |--------+---------+---------+---------+---------+---------+---------+---------+---------+---------+---------+---------+----|
       Nc  GDPRAAGEPSPQRV------GRR-YFSELSALEYFIIRHLAVVTMAPM-VEGEFTLVELIGLIETRKTPNFWKNIGKAFQ-KDG-RK---KKGVFGVPLEAIIERDGAESTDGVGPG-TLRIPAV
       Mg  --PRSHGEPSPQRGMVMGIGGRK-YFSELSNLDYFIVRHLAVLTMHPL-LNNEFELEELLGFIESRKPATFWNKFGKAFK-NDG-KKNVKKKGVFGVPLEMIIERDGADSTDGVGPG-TLRIPAI
       An  EQPYPAPTGR----------GKR-YFSELSALEYFIVRHVAVLSMEPL-LEGAFTLEELLSLIESRKP-TIWGIFGRAFT-KDA-KKGGKKKGVFGVSLDYLVEKEGTESTHGVGPG-ALRIPAL
       Af  ESHLAERTTK----------TKK-YFSELSALEYFIVRHVAVLSMEPL-LEGYFTLDELLSLIESRKP-TIWNIFGRAFN-KDA-KKAGKKKGVFGVSLDFLVEKEGTESTHGVGPG-ALRVPAL
       Sp  LTPKSEDAPI------------R-YLSDLSNLELLFTKHVAVLILHPL-VRDYYSLDELMEMSDLRKG-GFWEKFGKAFKGKDAEKKNVKKKGTFGVPLEILVERNNAQSTVGTGVG-VKHIPAF
       Ca  VLDNILNTSQPHPEEPVAVRKQK-YYSELTKDEHFIIRHIAVEALSHIS-KSYSNKEELLSLIQTKKQPTFWDKFKFG----GGDGKKDKVMAVFGVDLQVLTKKYGIDSDLGVGPS-KLRIPIV
       Um  IPHILEAEQAREQHRPLPSEGTR-CISELSALELFIVKHMAVMYLQQSALRDHVNLDDLIEFIETRKN-TFWGKIFKG----GKDKKEIKKKGVFGIPLEILVERNGADSTLGASAA-HLRVPSF
       Cn  DISLLADKDRRSDNRPL-------LSNLSPLQSIILRHFALLQLQKTSLGPLIDLDDMLDLLDARKG-QWWNKIFKG----NA-KKDTKKKGLFGVPIEVLVERTGSDSSFGVPNS-QVRVPEF
       Sc  GVYYSELSTM-----------------ELSMIRAISLSLLAGKQLISKTDPNYTSLVSMVFSNEKQVTGSFWNRMKIMM---SMEPKKPITKTVFGAPLDVLCEKWGVDSDLGVGPV-KIRIPII
       Cc  IPQLMEEAQAREQHRSLPRESSTPFIAELSPLELAIVKHFAVLQLSRSPLKDTFDLDEILEMVEMKKS-GFWNKLFN----KGDSKKALKKKGVFGVPLELLVEREGADSMLGASRGVTMRVPSF
Consensus  ........................s#Ls.l...i..h.Av..$..........l.......#..k...fW..............K..kkkgvFGvpl#.l.e..g.#S..G.......r!P..

           1376 1385      1395      1405      1415      1425      1435      1445      1455      1465      1475      1485      1495 1500
           |--------+---------+---------+---------+---------+---------+---------+---------+---------+---------+---------+---------+----|
       Nc  IDDILSLMRQM---DLSVEGVFRKNGNIKKLGELVEKIDKEG--NVDLSGNNVVQVAALLKRYLRELPDPLMTQKLYRLWLAAAK--ISDEDKKRQCLHLICCLMPKCHRDSMEILFCFLKWAGS
       Mg  VDDIVATMRQM---DLSVEGVFRKNGNIKKLSELCEKIDREGCDSVNLSSQPVVQVAALLKRYLRDLPDPLMTHKLYNLWLTAAK--IQDADKRRQCLHLICCLLPKCHRDCLEILFCFLKWAGT
       An  VDDSVSAMRQM---DMSVEGVFRKNGNIRRLKDISEMIDTKY-EQVDLTKENPVQIAALLKKFLREMPDPLLTFKLHRLFVASQK--IPDLEKQKRVLHLTCCLLPKAHRDTMEVLFAFLNWTSS
       Af  VDDAVSAMRQM---DMSVEGVFRKNGNIRRLKEISELIDNKY-DQVDLTKETPVQIAALLKKFLREMPDPLLTFKLHNLFVISQK--IPDPEKQKRLLHLTCCLLPKAHRDTMEVLFAFLNWTSS
       Sp  IGNTLAAMKRK---DMSVVGVFRKNGNIRRLKELSDMLDVSP-DSIDYEQETPIQLAALLKKFLRELPDPLLTFKLFGLFITSSK--LESEEERMRVLHLTICLLPKGHRDTMEVIFGFLYWVAS
       Ca  VDDIIAALRQK---DMSVEGIFRLNGNIKKLRELTEQINKNPLKSPDFSIQNAVQLAALMKKWLRELPNPLLTFGLYDMWVSSQR--QVNPVLRKRVLQLTYCMLPRSHRNLVEVLLYFFSWVAS
       Um  IDDVISAMKQM---DLSVEGIFRKNGNIRRLKELSEALDRDS-SAVNLLDDNPVQLAALLKKFLRELPDPLMTFKLHKLFVMSQK--LESEAERRRILHMVTILLPKGHRDTMEVLFAFLKWVAS
       Cn  IEDIISTMRQMGPSDMAVEGVFRKNGNIRKLQTLVEALDKDS-TAVNLSEENVIQLAALLKRFLRELPDPLLTFRLHKLFCAAAT--LPNLEERSRCLHLLVSLLPKYNRDTMEVLFIFLKWVSS
       Sc  IDELISSLRQM---DMSVEGIFRKNGNIRRLRELTANIDSNPTEAPDFSKENAIQLSALLKKFIRELPQPILSTDLYELWIKAAK--IDLEDEKQRVILLIYSLLPTYNRNLLEALLSFLHWTSS
       Cc  VDDVISAMRTM---DLSIEGIFRKNGNIRRLNDLKDAIDRDP-ASVDLTQDNPVQLAALLKKFLRDLPDPLLTFKLHRLFIASQSQNLPSDEDRKKRYLHLISLLLPRSHRDTMEVLFVFLKWVAS
Consensus  !d#..s.$rqm...D$s!eG!FRkNGNIr.L..l....#.......#...#n.!Q.aAL$Kkf1R#$P#Pl$tf.L..$...............r.lh$...$$P..hR#..Evlf.Fl.W..s

           1501 1510      1520      1530      1540      1550      1560      1570      1580      1590      1600      1610      1620 1625
           |--------+---------+---------+---------+---------+---------+---------+---------+---------+---------+---------+---------+----|
       Nc  FHQVDEESGSKMDIKNLATVIAPNVLFDRTATSTLGSDPMYAIEAVEVLIANIEEMCL------------------VPDELVQYISD--PAFLD--GELTTKEILKRYSERNTGGFGSRQY--A
       Mg  FHQVDDESGSKMDVKNLATVIAPNILYLQSKALTLDSDPMFAIVAVETLIQDIEEMCLISNKAMQ------------VPDDLVDILSD--STLFNNNGDLTTKEILKRFGDRPLMNGGLRGH--G
       An  FSHVDEETGSKMDIHNLATVMTPNILYPNTKTSAVD-ESFLAIEAVNALITYNDTMCE------------------IPEDLQSVLSD--SHLFKENSEVTTKDILKRYGDIARGGFSQKASNGG
       Af  FSHVDEDTGSKMDIHNLATVITPNILYPNTKNSTVD-ESFLAIEAVNALITYNDTMCEVNPSTTQCPVDQLADCLIKIPEDLQAVLSD--TTFFKDNNEVSTKEILKRYGDIARGSFSPKPNNGG
       Sp  FSHIDDEVGSKMDIHNLATVITPNILY--SKSNDPVDESFLAIEAVHLLIENFEKFCE------------------VPREISLLLDD--PTLFYNNAAWTSKELYKRCEEIISQMSLDERN---
       Ca  FSEIDEETGSKMDIHNLATVIAPNILI-SKQSSNSSSGNS-SNSGTNNSDSQQASGDN------------------YFLAIEVVNQL--IEQHEEFSIIPS-DIL---EFYEKCGFDKFDS---
       Um  FSHIDEETGSKMDLPNLATVICPNILY-SKGSDPTKDETFLANRAVLYLLEHQDEIWK------------------VPEELEAVLQD--KDLMNASSDLTSRDILKKCEKYAKLRQMRGGI---
       Cn  FSYREEETGSKMDLANLATVICPSILY-AKGSNAVKDDSFTAIAAVQELLENQDEYYR------------------VPSELLFVIQEGIYQIFAKDLDLPPKEIFRHCSKYTQARSAAQPQ---
       Sc  FSYIENEMGSKMDIHNLSTVITPNILYLRHKEISNDNVPDEPESGLVDSFAQNKGENYF-----------------LAIEIVDYL----ITHNEEMAMVPKFLMNLLKDVQLQKLDNYESINH
       Cc  FAHMDVEVGSKMDLPNLATVICPSILY-ARGRDALRDETFGALKVVTALLENQDEFFL------------------VPEEFLPVLQD--QEYFANAMELPSKEFLKKCDAYMRIKGGGRPT---
Consensus  F...#.#.GSKMD..NLaTVi.Pn!Ly..............a...v..l......................P.....l...............k...........................

           1626 1635      1645      1655      1665      1675      1685      1695      1705      1715      1725      1735      1745 1750
           |--------+---------+---------+---------+---------+---------+---------+---------+---------+---------+---------+---------+----|
       Nc  DVVEIVGRHEN-PSRPPPRRVESDPTSWQQETSVRTVTENAAYLGPGPTSGGGGGGGGGGSGGNNGGGMVDRTGGPGGAVSPQKRGGPGGGMGPHPLSQSHGDGGGPGGPQQQPIPYGPAPEHVA
       Mg  D--ELVGRHDT-PNRPFARRVDTDPAVWQDERSVRPVHDQQQYASSMHNTPLRRHEDNGQPSPY--GQDFDPSPGMSREPSSDRGGGQRREWRNSGWGNNSRQPGSVGVTGTG
       An  ETFTITN-QSRGANVPTSARIETDPYQDTASQMQSSVRHVPGPGGNSHPSGA--------IAPNKDLDKRTRSTSTGSQS-----EGQPQQV-PYRARPNPGPMGVAG
       Af  ETVTITNPHNRGANTPTSARIETDQSQDGPWQAQNPVRHVQNTGGHNHASSASAPYNGMELAPGQSASYRERSTSNGSQQNPIPQEGQPQQM-PYRSRPGAGPMGVAG
       Sp  ---TPKHTASTKRKRQPIRRVTTNLTSDVPSGSEVADTNSLSSTTKDEASPNSDAQPKPQVKPQHAVIRDS
       Ca  ---TKKEITTRDVMIKIDKELKEKPDYFDNFELKNP-TGSLTSQEVKRNSVSRIESKIQNRELN--GISER
       Um  ---SMGRNGSGGPLHAGELQNRHRPDVHPPTTMRQVHQREQMLHQQQQNAHSQYSSNNGALTPV--GTEFSRSPDKQGPWS----AGAVPPAGFSMSPTHRGPLGNQGHTPQKPLGYAQHGNQNG
       Cn  ---QQRLGIPHSNSSSQISQTSVSRSVESATRISQVSITAPDSTSGNYRPPLP
       Sc  FISTVMQSKTIDYSECDIKTPVTVKDSTTTVIQGEINK
       Cc  -----PGTPFNGPSGGSNGQPRYQQPLNSPTMERPPPIGMSASERSGRPSPNLSQHGHDSNHQNPNPNDYYQQQPQQQHQQ------QHQHQQHQQHPQHQQQQQPPSHSPSIPHANMQNLQSMS
Consensus  .............................................................................................................................

           1751 1760      1770      1780      1790      1800      1810      1820      1830 1835
           |--------+---------+---------+---------+---------+---------+---------+---------+----|
       Nc  AAALAGSGGGGPGVGGQGGNAVGVGGGDGAPGGGGGDYNRNGGSTTSGLHNHNGQQQQGGGPRKEWRNSGWGRPNNGVTTGTA
       Mg
       An
       Af
       Sp
       Ca
       Um  YAGQKPFYQAH--AQQQAQHMVRQPGSASATGLGPQAYNY
       Cn
       Sc
       Cc  RQPPMGVGNGGEQWNPNVQQLQPSQLSIAAPRPIAGNNHSMNSSRPSSVAGNRTPNAHDGRPPLDNQGSFYGSPNGYASPIRQRT
Consensus  ......................................................................................
```

Figure 10: Sequence alignment of *lrg-1* of various fungi.
Alignment of several fungal LRG1 proteins was done with Multalin (http://bioinfo.genopole-toulouse.prd.fr/multalin/; Um: *Ustilago maydis* (UM05343.1, only first 1600 aa were used for the alignment); Nc: *Neurospora crassa* (sequenced in this study, NCU02689); MG: *Magnaporthe grisea* (MGG_04377); An: *Aspergillus nidulans* (AN7576.2); Af: *Aspergillus fumigatus* (AFUA_2G15050); Sp: *Schizosaccharomyces pombe* (Rga1, SPBC3F6); Ca: *Candida albicans* (Ca019_7489); Cn: *Cryptococcus neoformans* (CNF02710); Sc: *Saccharomyces cerevisiae* (YDL240W); Cc: *Coprinopsis cinerea okayama* (CC1G_05680). Highly similar sequences (>94% identity) are shown in red, related sequences (70% to 94% identity) are shown in blue and less related sequences are shown in black. Whereas the fungal specific C-terminal RHO-GAP domain (red boxes) is conserved throughout fungi, the N-terminal, LIM domain containing part (green boxes) is shortened in the basidiomycete C*oprinopsis cinerea okayama*.

3.4.2 LIM and GAP domains of LRG1 are both essential for growth and septation

To analyze the function of the two LRG1 parts of *N. crassa* in more detail, constructs lacking either the LIM (LRG1$^{1\text{-}847}$) or the GAP motifs (LRG1$^{781\text{-}1279}$) were generated and expressed in *lrg-1(12-20)*. It was found, that both domains were necessary for the function of LRG1 (Figure 11). The importance of the GAP domain for the LRG1 function in hyphal morphogenesis was further supported by a construct, in which the conserved lysine 910 in the full-length protein was substituted with alanine. For p190GAP this mutation has been shown to result in a non-functional GAP due to loss of binding to the corresponding Rho1 (Li *et al.*, 1997). When this construct was either expressed in *lrg-1(12-20)* or in heterokaryotic *Δlrg-1*, no complementation of the mutant defects was detected (Figure 11). To confirm that the GAP activity is essential for the function of LRG1, an allele was generated, in which the conserved catalytic arginine residue 847 of the GAP was substituted by a lysine. This construct was expressed in *lrg-1(12-20)*, but was not sufficient for complementation (Figure 11) and provided strong evidence for the importance of the GAP activity for the function of LRG1. Sequencing of *lrg-1(12-20)* revealed a single amino acid substitution of the conserved tyrosine 926 in the GAP domain by histidine (TAC to CAC), supporting that the GAP domain is essential for the cellular function of LRG1.

Functions for the LIM domains were implicated by the failure of LRG1$^{781\text{-}1279}$ to complement *lrg-1(12-20)*. Therefore, a series of *lrg-1* alleles were generated, in which 4 of the 8 metal-ion-coordinating cysteines or histidines of each of the three LIM domains were substituted mainly by serines (as illustrated as yellow residues in Figure 11 C; for LIM domain 1: C121S; C124S; C98L; C101S; for LIM domain 2: H185V; C188S; C162S; C165A; for LIM domain 3: C492S; C495S; C469G; C472S). Alanine and serine residues have been shown to mimic the behaviour of cysteines, but lack their zinc-coordinating activity, which has been shown to be essential for the function of LIM domains (Bombarda *et al.*, 2002; Schmeichel and Beckerle, 1997).

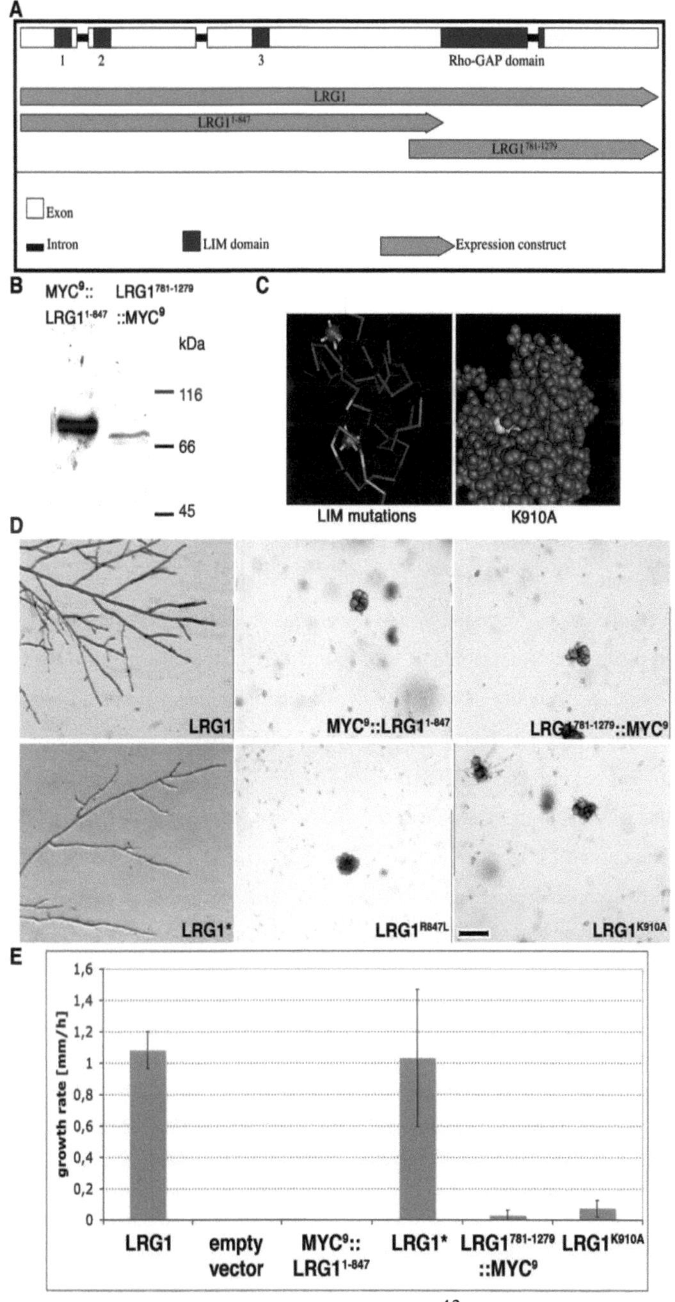
A
1
2
3
Rho-GAP domain
LRG1
LRG1 1-847
LRG1 781-1279
Exon
Intron
LIM domain
Expression construct
B
MYC9::
LRG1 1-847
LRG1 781-1279
::MYC9
kDa
116
66
45
C
LIM mutations
K910A
D
LRG1
MYC9::LRG1 1-847
LRG1 781-1279::MYC9
LRG1*
LRG1 R847L
LRG1 K910A
E
growth rate [mm/h]
1,6
1,4
1,2
1
0,8
0,6
0,4
0,2
0
LRG1
empty vector
MYC9:: LRG1 1-847
LRG1*
LRG1 781-1279 ::MYC9
LRG1 K910A

Figure 11: The N-terminus and the C-terminal Rho-GAP domain are both required for cellular functions of LRG1.
(A) Genomic organization of *lrg-1* locus. The four exons are represented as white blocks and the three introns as black bars. The *lrg-1* gene encodes for three LIM domains (brown) in the N-terminal half and one Rho-GAP domain (blue) in the C-terminal half of the protein. Green arrows represent expression constructs under the control of the endogenous promoter used for complementation experiments. The MYC^9 tag of the tagged constructs was fused to the N-terminus of the LIM and the full length LRG1. The MYC^9 tag was fused to the C-terminus of the GAP and the full-length LRG1 construct. (B) Western blot of protein extracts from MYC^9::LRG1$^{1\text{-}847}$ and LRG1$^{781\text{-}1279}$::MYC^9 strains probed with anti-MYC antibody confirmed the expression of the MYC^9 tagged deletion constructs. (C) Illustration of point mutations brought in the full-length protein. Conserved residues in the predicted structural domains that are mutated in the corresponding complementation constructs are indicated in yellow (for detailed information on mutated residues see text). (D) Colony morphology of *lrg-1(12-20)* transformed with the indicated constructs. Transformants were grown at 37°C to determine the complementation capability of the respective construct. Transformations with the corresponding untagged constructs resulted in the same phenotype as for the tagged constructs (not shown). Scale bar is 200 μm. (E) Quantification of growth rates of the indicated strains (standard deviation is indicated, n=3). LRG1*: construct harbouring multiple LIM point mutations (see text for details).

Surprisingly, when the three constructs, each of which was defective in one of the three LIM domains, were tested for their ability to restore growth, they did lead to full phenotypic complementation of *lrg-1(12-20)*. Also, when the three mutated LIM domains were combined to generate the triple LIM domain mutant LRG1*, no abnormal growth behaviour of *lrg-1(12-20)* (Figure 11 D, E) or homocaryotic *Δlrg-1* (data not shown) transformants were observed, and overall radial growth rates on agar plates were comparable to *wild type* (*LRG1*: 1,1 +/- 0,1 mm/h; *LRG1**: 1,0+/- 0,4 mm/h; n = 3). A higher variability of extension rates of individual hyphal tips was noted for *LRG1** grown on microscopic slides (9,2 +/- 8,2 μm/min; n = 50) in comparison to *LRG1* (6,3 +/- 5,3 μm/min; n = 50), suggesting that the dysfunctional LIM domains impair proper tip extension. Nevertheless, the placement of septae was not altered in *LRG1**. The distance of the hyphal tip to the first septum was 224 +/- 41,9 μm in *LRG1* and 225,4 +/- 55,8 μm for *LRG1** (n=30), while the distance between subapical septae was 84,9 +/- 30,9 μm in *LRG1* compared to 92,5 +/- 32,4 μm in *LRG1** (n=100), indicating that the LIM domains are dispensable for normal growth, in contrast to the entire N-terminus of LRG1.

3.4.3 The LIM domains are required for localizing LRG1 to sites of growth

To determine the cellular distribution of LRG1, N-terminal and C-terminal *myc*^{9}-tagged versions of *lrg-1* were generated, which both complemented the *lrg-1(12-20)* growth defects. Immunolocalization experiments using anti-MYC monoclonal antibodies revealed a punctated distribution of MYC^9::LRG1 throughout the cell, which was enriched at hyphal tips and along septae (Figure 12 A, data for septae not shown).

This localization was in line with a function of LRG1 at growing tips and was also observed, when the generated and purified antibody against LRG1 was used to detect the endogenous LRG1 in *wild type* hyphae (data not shown). To determine which region of LRG1 is responsible for its tip-enriched localization, the cellular distribution of MYC^9::$LRG1^{1-847}$ and $LRG1^{781-1279}$::MYC^9 was determined in the *wild type* background. While $LRG1^{781-1279}$::MYC^9 aberrantly localized in a cortical fashion throughout the whole hypha and along septae, the MYC^9::$LRG1^{1-847}$ distribution was similar to the localization of LRG1, suggesting that the N-terminal part of LRG1 may be involved in the localization of LRG1.

For a better resolution of the dynamics of LRG1, N-terminal GFP fusion constructs with LRG1 and with LRG1* (containing the three mutated LIM domains) were generated. Both constructs were inserted into *Δlrg-1* and complemented the growth defect. ApoTome based fluorescence microscopy confirmed the apical vesicular-reticulate localization of GFP::LRG1 and resulted in a more diffuse localization of GFP::LRG1* lacking the prominent streaks (Figure 12 B). In addition to this tip-enriched vesicular-reticulate localization, a GFP::LRG1 cap along the apical cortex and strong staining around the septal pore was frequently observed (Figure 12 C). In contrast, GFP::LRG1* was distributed in a more diffuse manner throughout the hypha with only a weak accumulation at the hyphal apex and no localization at septal pores. Thus, both the immunolocalization data and the GFP-fusion proteins indicated a function of the LIM domains in localizing the protein with its GAP domain to sites of active growth in the apical region and along septae. To test whether the GFP::LRG1 stained streaks may be any membranous compartments of the cell, colocalization experiments with the lipophilic life dyes ER-Tracker (staining the endoplasmatic reticulum), FM4-64 (staining endosomes) and Mito-Tracker (staining mitochondria) were performed. In all three cases, no obvious colocalisation was observed (data not shown).

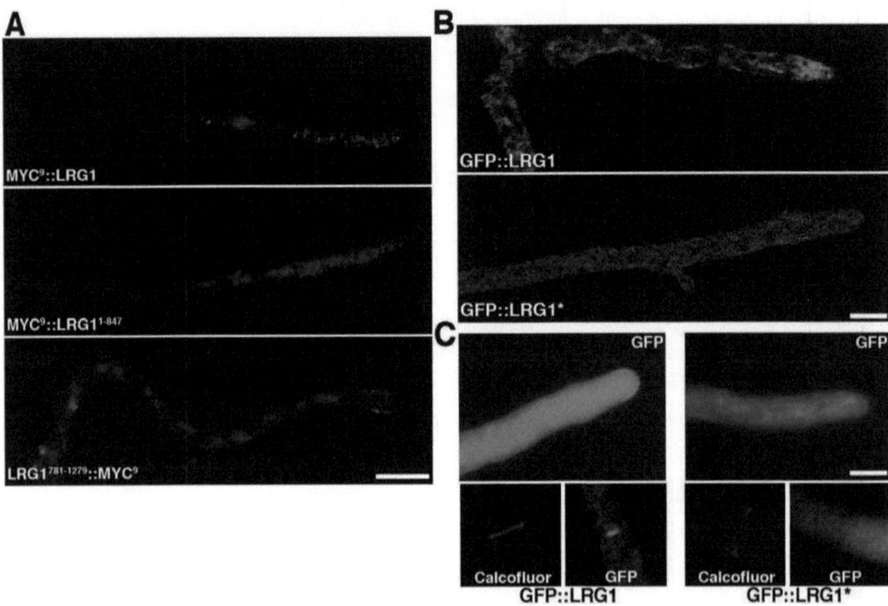

Figure 12: LRG1 localization to sites of growth at the hyphal tip and the septum is dependent on functional LIM domains.

(A) Immunolocalization of the indicated strains using anti-MYC antibodies. MYC^9::LRG1 and MYC^9::$LRG1^{1\text{-}847}$ localized in vesicular-reticulate structures that are enriched at hyphal tips, while $LRG1^{781\text{-}1279}$::MYC^9 miss-localized at the hyphal cortex all along the hypha. Scale bar is 10 μm. (B) Localization data generated in the ApoTome modus confirmed the apical vesicular-reticulate localization of GFP::LRG1 and showed a more diffuse localization of GFP::LRG1* lacking the prominent streaks. (C) Standard fluorescence microscopy was used to deternine the localization of the indicated GFP tagged proteins. GFP::LRG1 localized as a crescent at growing hyphal tips and at septae by (left panel), while GFP::LRG1* containing the three mutated LIM domains did not predominantly localize at hyphal tips and septae (right panel). Scale bar for (B) and (C) is 5μm.

To characterize the dynamics of GFP::LRG1 in more detail, I first asked, if the localization of the apical cap is growth-dependent and analysed 100 randomly chosen tips of the GFP::LRG1 expressing strain (Figure 13 A). 70% grew at rates of ≥ 0.2 μm/sec and displayed an apical cap, while 2% grew at these rates and did not show a cap. In contrast, 10% of the hyphae grew at rates of < 0.2 μm/sec and displayed an apical cap, while 18% grew at this rate and did not show a cap. Thus, the localization of GFP::LRG1 as a cap-like structure at the hyphal tip was dependent on active growth. A different behaviour was observed, when the localization of GFP::LRG1* was quantified. Only 13% of the hyphal tips grew fast and 2% grew slow and displayed a cap, while 62% and 23%, respectively, did not, confirming that the LIM domains are responsible for localizing LRG1 to the growing hyphal apex. During this analysis a correlation between the size of the apical cap and the growth rate was noted (Figure 13 B).

GFP::LRG1 accumulated primarily along the central apical cortex in slow growing hyphae, while in fast growing tips, an extended cap structure was visible with the highest GFP intensity frequently accumulating as a subapical ring approximately 2 μm behind the apex. When the length of the apical cortex GFP::LRG1 was quantified relative to the growth rate of the tip, a positive correlation ($R^2 = 0,6204$) between growth rate and size of the apical cap was observed (Figure 13 B).

Septum formation requires the formation of actin rings as an initial step prior to cross wall formation. Later, active glucan and chitin synthesis (visualized by Calcofluor White staining) led to the completion of septation (Harris, 2001; Rasmussen and Glass, 2005, 2007). For the understanding of LRG1's function during septum formation, the dynamics of GFP::LRG1 in subapical regions was analysed (Figure 14 C). Calcofluor White positive cross walls appeared prior to the accumulation of weak GFP::LRG1 dots along the cortical region of a forming septum. These dots merged into a septal plate, resulting in weak labeling of the whole septum after approximately 3 min. This diffuse septal localization then disappeared for about 10 min before GFP::LRG1 strongly accumulated in the central region around the septal pore. This final localization remained permanent and was visible at most older septae. When growing hyphae were incubated with 1 μM latrunculin A (Figure 14 D), the apical cap disappeared within 1 min, prior to any morphological change that was induced by the actin-depolymerizing agent after about 5 min. Nevertheless, the septal pore-associated localization of GFP::LRG1 remained unaffected by this treatment, indicating that this terminal association of LRG1 with the central region of the septum is independent of a functional actin cytoskeleton. To test, whether the localization of LRG1 depends of on a functional microtubule cytoskeleton, the *GFP::LRG1* strain was incubated with 4 μg/ml nocodazole or 10 μg/ml benomyl (both drugs are microtubule depolymerizing agents). For both treatments, the apical GFP::LRG1 cap persisted as long as growth was observed, indicating that a functional microtubule cytoskeleton is dispensable for LRG1's localization (data not shown). However, the size of the apical GFP::LRG1 cap was reduced in nocodazole treated cells when compared to untreated tips growing at comparable rates (Figure 13 E), suggesting that a functional microtubule cytoskeleton may be necessary for stable tip localisation of LRG1.

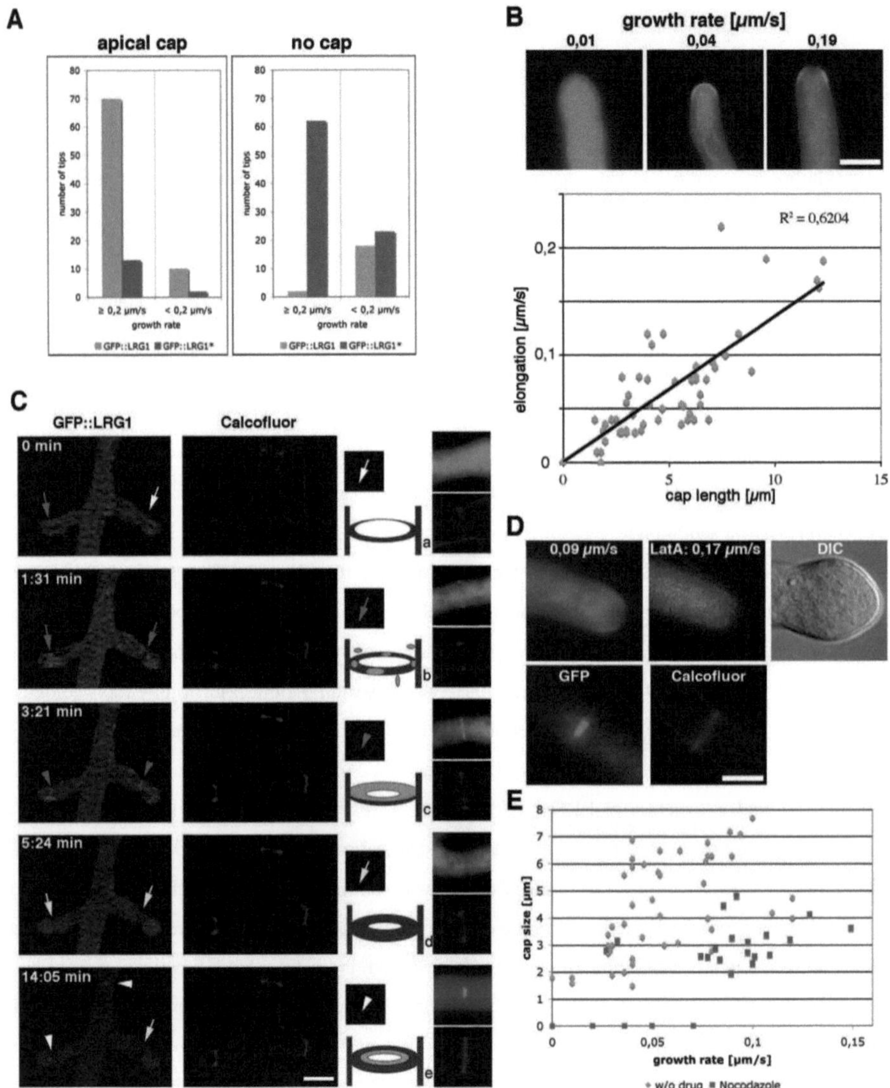

Figure 13: The localization of GFP::LRG1 is dependent on active growth and functional LIM domains.

(A) The dependence of the presence of an apical GFP::LRG1 or GFP::LRG1* cap from the growth rate of 100 randomly chosen hyphal tips was assayed. (B) The cap size and growth rate were determined by measurement of the distance of the apical cortex of hyphal tips between two images that were taken with a 20 s time-lapse automatisation. The size of the apical LRG1 cap correlated with the growth rate of the tip. The growth rate was plotted against cap length. Examples of slow, medium and fast growing tips are shown in the upper panel. Scale bar is 5 μm.

Figure 13, continued: (C) To determine the septum localization, images of a few time points were taken. GFP staining was analysed for a short time every minute to reduce bleaching. When a change in GFP localization was visible, pictures were taken. The absolute time points of the GFP images were used for the calculation of relative time points. The GFP::LRG1 localization at the forming septum occurs in two phases. Approximately 1.5 min after the appearance of Calcofluor White positive septae (a, white arrows) GFP::LRG1 began to localize along the septal plate initially as punctate specs (b; red arrow) that soon merged into a diffuse plate (c; red arrow heads). This septal GFP::LRG1 localization disappeared then for ca. 10 min (d; yellow arrow), before it strongly accumulated at the central part of the septum along the septal pore (e; white arrow heads). Examples of the stages are shown in the left panel for better visualisation. Scale bar is 5 μm. (D) Treatment of the GFP::LRG1 expressing strain with 1 μM latrunculin A affected the localization of GFP::LRG1 at the hyphal apex within 30 s (upper panel). This was prior to any morphological change, which occurred within 5 min (DIC image). The treatment had no influence on the terminal localization of GFP::LRG1 along the septal pore (lower panel). Scale bar is 5 μm. (E) 60 μl of 6 μM nocodazole were added to hyphae grown on agar blocks of approximately 30 μl volume. 2 min later still growing tips were analysed as in (B). For growth rates up to 0.15 μm/s the cap size of GFP::LRG1 was plotted against the growth rate for untreaded and nocodazole treaded cells. The average cap size at comparable growth rates is reduced.

3.4.4 LRG1 is a RHO1 specific GAP

To determine, which of the six RHO GTPases encoded in the *N. crassa* genome are regulated by the Rho-GAP domain of LRG1, the *wild type* forms of all RHO GTPases were ectopically expressed in *lrg-1(12-20)* under the control of a modified, constitutively active *cpc-1* promoter. Only RHO1 was capable of partially complementing the *lrg-1(12-20)* growth defect (Figure 14 A). Also dominant active (G15V) as well as dominant negative (E41I) RHO1 alleles were likewise able to initially overcome the *lrg-1(12-20)* dependent growth defect (Figure 14 A). It should be noted that all RHO1 transformants died within 2 days with swollen and lysed hyphae (Figure 14 B). This lethal phenotype was also observed when the three RHO1 alleles were overexpressed in *wild type* and is probably the result of interfering with endogenous RHO1 function.

Table 5: Exon-intron boundaries for RHO1, RAC and CDC42 coding regions.

Rho protein/ NCU number	Chromosom / Strand	Exon number	Start	End	Size (bp)
RHO1/NCU01484.3	II / (-)	1	32847	32777	70
		2	32640	32613	27
		3	32348	32151	197
		4	32083	31945	138
		5	31843	31688	155
RAC/NCU02160.3	I / (-)	1	1211668	1211636	32
		2	1211476	1211444	32
		3	1211367	1211308	59
		4	1211226	1211046	180
		5	1210971	1210739	232
		6	1210659	1210596	63
CDC42/NCU06454.3	III / (-)	1	192396	192377	19
		2	192182	192063	119
		3	191926	191548	378
		4	191475	191404	71

To investigate the activity of LRG1 cDNAs of the six Rho genes encoded in the genome of *N. crassa* were amplified. The sequencing of the cDNA amplicons differed from the *in silico* predictions for *rho-1*, *rac* and *cdc42*. The obtained exon-intron boundaries are shown in Table 5.

The coding sequences were first expressed from pPicholi-C derived plasmids in *Pichia pastoris*. This led to low expression and inefficient purification of RHO proteins, possibly due to toxic effects. Also bacterial expressed tagged *N. crassa* Rho proteins were expressed in low amounts and purified in low abundance (compared to purification of human forms). Maltose binding protein (MBP) (data not shown) or glutathione S-transferase (GST) fusions (Figure 14 C) of the six Rho GTPases were purified to an extent that was sufficient for *in vitro* activity studies.

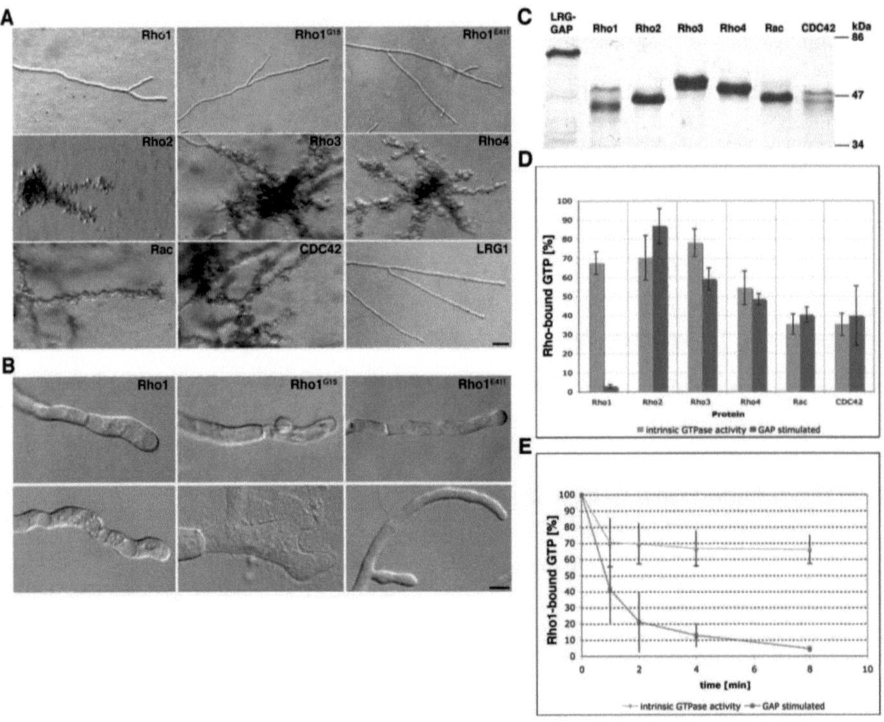

Figure 14: LRG1 is a specific GTPase activating protein for Rho1.
(A) Overexpression of *wild type* as well as dominant-active (Rho1^{G15V}) and dominant-negative (Rho1^{E41I}) RHO1 alleles, but not of any of the other Rho proteins was sufficient to temporarily overcome the morphological defect of the *lrg-1(12-20)* mutant at restrictive temperature. (B) Morphology is shown of *wild type* cells (upper panel) or *lrg-1(12-20)* cells (lower panel) transformed with the overexpression constructs of the indicated RHO1 alleles. The cells died within two days. Scale bar is 10 μm. (C) Coomassie staining of bacterially expressed and affinity-purified GST::RHO proteins and of the GST tagged Rho-GAP domain used for the *in vitro* GAP assays (D) Intrinsic and GST::LRG1$^{650\text{-}1035}$ stimulated GTPase activites of the six

Rho proteins indicated that LRG1 is a RHO1 specific GAP. (E) Kinetics of *in vitro* RHO1 GTPase activity in the presence and absence of GST::LRG1$^{650\text{-}1035}$.

Next, *in vitro* GTPase assays were performed with the six bacterially expressed His6::GST::Rho fusion proteins (Figure 14 D, E). All G-proteins had significant and unexpectedly high intrinsic GTP-GDP turnover rates, indicating that the enzymes were active.

The addition of GST:: LRG1$^{650\text{-}1035}$, the GAP domain of LRG1, stimulated only the GTPase activity of RHO1 (approximatly ten fold). None of the other GTPases caused any significant change in RHO activity (a significant GTPase stimulation means at least two fold higher activity). These data are consistent with the genetic analysis and identify LRG1 as a RHO1 specific GAP.

3.4.5 LRG1 regulates several output pathways of RHO1

RHO1 is a *bona fide* regulatory subunit of the cell wall enzyme β1,3-glucan synthase. Furthermore, it activates PKC and subsequently the cell wall integrity MAP kinase pathway, which in turn regulates the transcription of several cell wall specific enzymes. In addition, RHO1 regulates the actin cytoskeleton by directly activating the formin BNI1 (Beauvais *et al.*, 2001; Evangelista *et al.*, 2003; Levin, 2005; Sharpless and Harris, 2002). Therefore, the impact of defective LRG1 function on each of these signalling pathways was tested.

The compensation of the *lrg-1(12-20)* growth defect by the osmotic stabilization of the medium with 1.75 M sorbitol or 1 M NaCl (Figure 15 A) and the abnormal chitin distribution detected by Calcofluor White staining (Figure 4 D) indicated an altered cell wall structure.

Predicted genetic interactions between *lrg-1* and *gs-1* were analysed due to the involvement of RHO1 as activating regulatory subunit of the glucan synthase (GS). The mutant screen in *N. crassa* had resulted in the isolation of a temperature sensitive glucan synthase mutant strain called *gs-1(8-6)* (Seiler and Plamann, 2003). A *lrg-1(12-20);gs-1(8-6)* double mutant displayed strong synthetic defects (Figure 15 B).

Under restrictive conditions, *gs-1(8-6)* generated growth-inhibited hyperbranched hyphae that were still capable of slow apical growth (Seiler and Plamann, 2003). *lrg-1(12-20);gs-1(8-6)* generated in contrast apolarly growing spheres when germinated at 37°C. When the double mutant was germinated at permissive conditions and transferred to restrictive temperature, chains of spherical growing cells were generated within 10 h. The growth rate of *gs-1(8-6), lrg-1(12-20)* at permissive or semi-restrictive temperature was reduced by 25% and 41%, respectively, when compared to *lrg-1(12-20)* as the slower growing of the two parental strains.

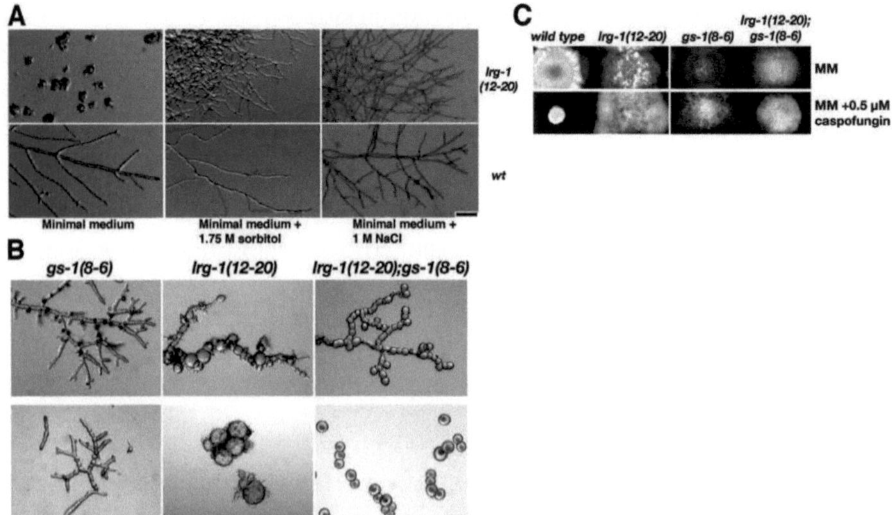

Figure 15: LRG1 affects GS1-dependent cell wall functions.
(A) Osmotic stabilization of *lrg-1(12-20)* on medium containing NaCl or sorbitol results at restrictive temperature in a suppression of the growth defect. Scale bar is 200 µm. (B) The indicated temperature-sensitive strains were grown at permissive temperature and shifted to 37°C for 10 h (upper panel) or germinated at restrictive temperature for 15 h (lower panel). The double mutant shows strong synthetic defects. Scale bar is 40 µm. (C) *lrg-1(12-20), gs-1(8-6)* and *lrg-1(12-20);gs-1(8-6)* are hyposensitive against the β1,3-glucan synthase inhibitor caspofungin.

Furthermore, these strains were less sensitive to Caspofungin (Figure 15 C), a specific inhibitor of fungal β1,3-glucan synthesis (Denning, 2003), indicating increased β1,3-glucan synthase activity in *lrg-1(12-20)*, *gs-1(8-6)* and *lrg-1(12-20);gs-1(8-6)*. Thus, hyperactive RHO1 in *lrg-1(12-20)* is impairing the cell wall organization of *N. crassa* by increasing β1,3-glucan formation, and the *lrg-1(12-20);gs-1(12-20)* double mutant defect is likely due to combined GS1 hyperactivities of the two strains.

The abnormal chitin distribution observed in *lrg-1(12-20)* (Figure 4 D) could be a result of increased activity of the *N. crassa* PKC/MAK1 cell integrity pathway, which would subsequently result in the altered expression of additional cell wall enzymes. Therefore, the activity level of PKC was altered in the *lrg-1* mutant and in *wild type*. A partial suppression of the *lrg-1(12-20)* growth defect on media supplemented with 1 μM staurosporine, a protein kinase inhibitor with highest specificity towards PKC was observed (Figure 16 A). Nevertheless, staurosporine is also inhibiting protein kinase A (although with a about ten fold lower affinity). Therefore, the impact of 50 μM KT5720, a PKA specific inhibitor, was tested on the growth behaviour of *lrg-1(12-20)*, but no morphological changes were observed, suggesting that PKC is hyperactive in *lrg-1(12-20)*. This was further supported by growth tests on media supplemented with 60 nM cercosporamide, a new and highly specific inhibitor for PKC (Sussman *et al.*, 2004).

Next, phospho-specific antibodies against activated ERK-type MAPKs were used to determine the phosphorylation of the cell integrity MAP kinase MAK1, which correlates with its activity. No obvious differences were found in MAK1 phosphorylation pattern in *wild type* extracts from cultures grown at 25°C and 37°C. Stressing these cells by a temperature shift from 25°C to 37°C for 20 min resulted in activation of cell integrity signalling and increased MAK1 phosphorylation (Figure 16 B). In contrast, the MAK1 phosphorylation pattern in *lrg-1(12-20)* shifted to restrictive temperature for 20 min did not result in heat stress-induced phosphorylation of MAK1, suggesting that MAK1 regulation may be affected in *lrg-1(12-20)*. However, an increase in MAK1 activity levels would be predicted for increased RHO1 – MAK1 signalling. No increase in MAK1 phosphorylation levels were detected in *lrg-1(12-20)* after prolonged temperature shift (Figure 16 B) or in *Δlrg-1* germinated on benomyl and panthotenic acid (data not shown). To test, whether the capacity of the MAK1 pathway to respond to stress signals is affected in *lrg-1(12-20)* in a general manner, *wild type* and *lrg-1(12-20)* were grown at 37°C and stressed by the addition of 7 mM H_2O_2 for 20 min (Figure 16 C). Both strains displayed identical activation patterns as detected by MAK1 phosphorylation, indicating that the response capacity of MAK1 is not affected in *lrg-1(12-20)*. Taken together, these data indicate that RHO1 – PKC signalling is increased in *lrg-1(12-20)*, but also suggest that the downstream MAK1 MAP kinase pathway is only affected to a minor extent.

An altered organization of the actin cytoskeleton in *lrg-1(12-20)* became evident by the hypersensitivity of the mutant to the actin depolymerising drug latrunculin A (Figure 16 D). BNI1 is the exclusive formin encoded from the *N. crassa* genome and contains a C-terminal effector domain and a N-terminal GTPase-binding domain (GBD) that mediates binding to RHO1. To further support the impact of *lrg-1(12-20)* on RHO1 dependent actin polymerization, the N-terminal half (amino acids 1-824) of BNI1 was overexpressed in *lrg-1(12-20)* (Figure 16 E). This construct has been shown to bind to activated RHO1 and is acting in a dominant-negative fashion by quenching activated RHO1 (Alberts, 2001; Palazzo *et al.*, 2001).

The overexpression of $BNI1^{1-824}$ resulted in partial suppression of the *lrg-1(12-20)* defects as detected by the formation of up to 0.5 cm large colonies at restrictive temperatures in 2 days, compared to *lrg-1(12-20)* transformed with the hygromycin resistance marker that reached a maximal colony diameter of 0.1 cm.

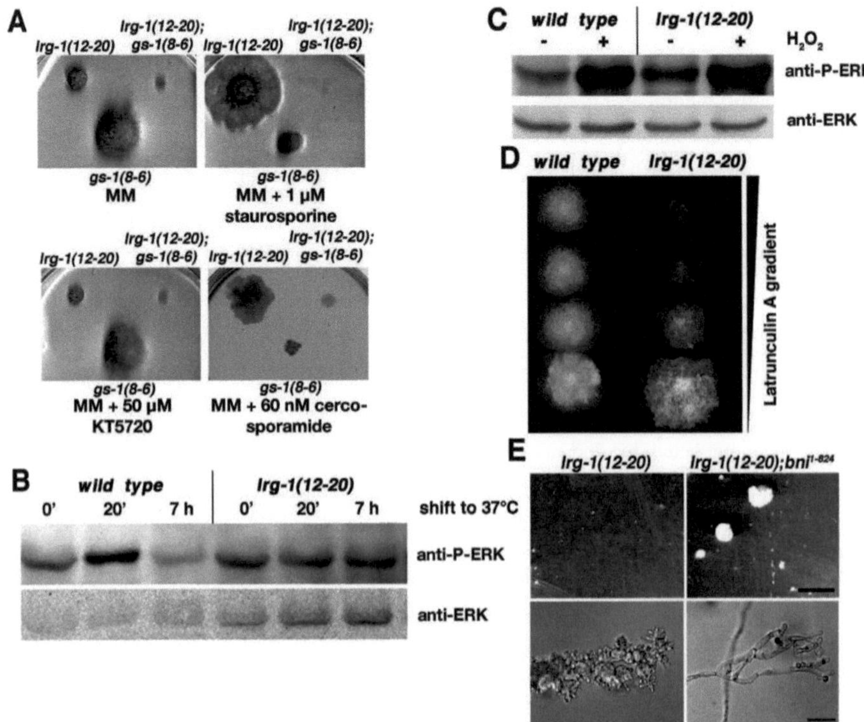

Figure 16: The MAK1 cell integrity pathway and formin regulation are affected in *lrg-1* mutants.
(A) The indicated strains were grown either for 40 h at 32°C on minimal media (MM) plates or MM plated supplemented with the kinase inhibitors staurosporine or KT5720 or for 6 days on plates supplemented with cercosporamide. The growth defects of *lrg-1(12-20)* were suppressed by staurosporine and cercosporamide, but not by KT5720. (B) Total soluble protein was extracted from *wild type* and *lrg-1(12-20)* shifted to 37°C for the indicated times. The blot was probed with anti-phospho-ERK (anti-P-ERK) antibody to detect activated MAK1 (upper panel). A replicate was probed with anti-ERK to confirm equal loading (lower panel). *wild type* but not *lrg-1(12-20)* showed stress induced MAK1 activation after 20 min at 37°C. (C) MAK1 phosphorylation induced by the addition of 7 mM H_2O_2 for 20 min to *wild type* and *lrg-1(12-20)* indicate that the response capacity of the MAK1 pathway is not affected in *lrg-1(12-20)*. (D) *lrg-1(12-20)* is hypersensitive against the actin depolymerising drug latrunculin A. (E) The increased colony size of $BNI1^{1-824}$ transformants of *lrg-1(12-20)* grown at 37°C for 2 days on agar plates (upper panel; Scale bar is 0.5 cm) and the enhanced tip extension and reduced hyperbranching of the strains (lower panel; Scale bar is 30 µm) indicate a partial rescue of *lrg-1(12-20)* by overexpression of $BNI1^{1-824}$.

A microscopic analysis of these colonies revealed, that the leading hyphae generated dome shaped tips. The tips were capable of apical extension, although hyperbranching still occurred.

3.5 LRG1 acts parallel with the Ndr kinase COT1 in a motor protein dependent manner

The morphological defects of *lrg-1(12-20)* and of Δ*lrg-1* were strikingly similar to those of the Ndr kinase mutant *cot-1* and its upstream regulating kinase *pod-6* (Seiler *et al*., 2006; Yarden *et al*., 1992) and allowed to hypothesize that LRG1 and the COT1 and POD6 including pathway may have related functions. The common defects of these mutants are a block in apical tip extension and the generation pointed/needle-like tips, the induction of multiple subapical branches, excessive and mislocalized chitin distribution and a highly thickened cell wall (see section 3.1). In contrast to a *cot-1(1);pod-6(31-21)* double mutant, which displayed defects identical to the two parental strains, *cot-1(1);lrg-1(12-20)* or *pod-6(31-21);lrg-1(12-20)* double mutants were synthetically lethal (Figure 17 A and data not shown). Double mutant ascospores of both, *pod-6; lrg-1* or *cot-1;lrg-1*, which were germinated at permissive conditions, generated polar germ tubes, but grew in a compact and highly vacuolated manner with very slow tip extension rates and frequent lysis of hyphal tips or subapical regions. Germination at 37°C resulted in apolar germination and death of the double mutants, indicating two parallel, a COT1/POD6- and an LRG1-dependent, pathways for apical tip extension.

Mutations in the dynein/dynactin complex (called *ropy* for their cork-screw growth phenotype) partially suppressed the *cot*-1 and *pod-6* defects by a reduced retrograde transport rate of vesicles and subsequent accumulation of kinase in apical areas (Bruno *et al*., 1996; Gorovits and Yarden, 2003; Seiler *et al*., 2006). Given the genetic connection between *lrg-1* and *cot-1*, both mutants may also share the same suppressors. Thus, growth rates of various *ropy* mutants as single as well as double mutants in combination with *lrg-1(12-20)* and *cot-1(1)* grown at restrictive temperatures were compared, and it was found that the dynein/dynactin mutations partially suppressed the *lrg-1(12-20)* phenotype in a manner identical to *cot-1(1)* (Figure 17 B). In contrast, *gul-1*, a mutant that had been implicated in phosphatase-associated, COT1-antagonizing functions (Seiler *et al*., 2006; Terenzi and Reissig, 1967), was suppressing only *cot-1(1)* and *pod-6(31-21)*, but not *lrg-1(12-20)*.

Next a genetic connection between *lrg-1(12-20)* and *nkin*, which encodes for the microtubule plus-end directed motor protein conventional kinesin was investigated. Kinesin has been shown to counteract dynein's retrograde transport activity. It has also been implicated in the localization of COT1 and POD6 (Seiler *et al*., 1999; Seiler *et al*., 2006; Zhang *et al*., 2003) and may be involved in distribution of regulatory components. The growth rate of a *lrg-1(12-20);nkin*^{RIP} double mutant at semi-restrictive temperature was reduced by 42% when compared to *lrg-1(12-20)* as the slower growing of the two parental strains, indicating a synthetic interaction between *lrg-1* and *nkin*. This suggests that both opposing microtubule-dependent motor proteins may be involved in localizing LRG1 similar to what has been reported for COT1 and POD6 (Seiler *et al.*, 2006).

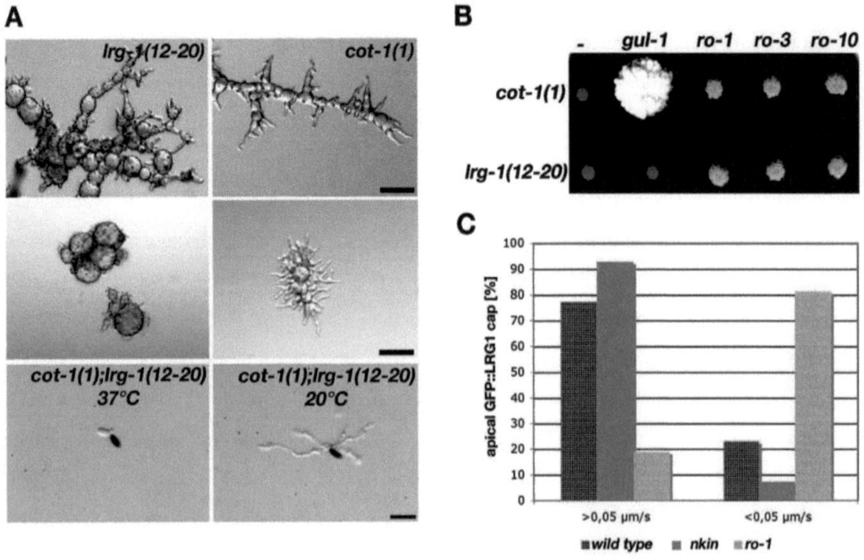

Figure 17: LRG1 acts in a pathway parallel to the Ndr kinase COT1 and is localized in a motor protein dependent manner

(A) *lrg-1(12-20)* and *cot-1(1)* displayed similar defects with tip extension terminating with 2-10 and 10-30 μm long pointed tips, respectively, when shifted from permissive to restrictive conditions for 12 h (upper panel) or when germinated at 37°C for 15 h (middle panel). Double mutant ascospores germinated at 20°C produced compact, highly vacuolated and slow growing hyphae. Germination at 37°C resulted in apolar germination and death of *cot-1(1);lrg-1(12-20)* ascospores. (lower panel). Scale bar is 40 μm. (B) *cot-1(1), lrg-1(12-20)* and the respective double mutants with components defective in the phosphatase associated protein *gul-1* or in the dynein/dynactin subunits *ro-1, ro-3,* and *ro-10* were cultivated for 3 days at 37°C. The increased colony diameter of *lrg-1(12-20);ro-1/-3/-10* mutants compared to *lrg-1(12-20)* indicates suppression of the GAP defect by loss of dynein/dynactin function. (C) Apical GFP::LRG1 caps were detected in 78% of *wild type* tips with growth rates of >0,05 μm/s. This rate of GFP::LRG1 caps increased in mutants defective in the anterograde directed motor protein *nkin* to 91% and decreased to 19% in the mutants defective in the retrograde directed motor protein *ro-1*.

Therefore, the localization of GFP-tagged LRG1 in these mutant backgrounds was determined. As shown previously (Figure 13 A), the majority of slow growing hyphal tips of *wild type* displayed no GFP::LRG1 cap. The growth rates of the two motor mutants is about 20% of *wild type* (Seiler *et al.*, 1997; Seiler *et al.*, 1999), and thus, a visible cap should be rarely visible, if the GFP::LRG1 localization is independent of the two motors. Therefore, the growth rate of hyphal tips which showed an apical GFP::LRG1 cap in these backgrounds was determined. 22% of wild type, while 81% of dynein heavy chain mutant (*ro-1)* tips with growth rates <0,05 μm/s displayed an apical GFP::LRG1 cap and the number of GFP::LRG1 caps at these slow growth rates was reduced to 8% in the *nkin* background (Figure 17 C). Thus, defective retrograde transport in the dynein heavy chain mutant resulted in increased apical localization of GFP::LRG1, while reduced forward directed transport inhibited the apical accumulation of GFP::LRG1, indicating that the apical localization of LRG1 is dependent on opposing microtubule-dependent motor proteins.

These results may be explained by a connection between COT1 and RHO1 signalling. Therefore, the RHO1 effector pathways that are affected by LRG1 were analysed in *cot-1(1)*. A highly thickened cell wall of *cot-1(1)* grown at 37°C was described in electron microscopic sections (Collinge *et al.*, 1978; Gorovits *et al.*, 2000) and is also found in *pod-6(31-21)* (Figure 9). In addition, the Calcofluor White label in temperature shifted *cot-1* hyphae was similar to the excessive chitin distribution at sites of aberrant growth in *lrg-1(12-20)*, which already indicated a defect in cell wall organization. However, no synthetic interaction in a *cot-1(1);gs-1(8-6)* double mutant was observed (Figure 18 A) and growth characteristics of *cot-1(1)* were identical to *wild type* on media supplemented with caspofungin (Figure 18 B). Thus, these data suggest an effect of COT1 on the cell wall, but no direct regulation of GS1 by COT1.

Therefore, the activity of the cell wall integrity MAP kinase pathway was analysed. Phospho-specific antibodies detected induced MAK1 phosphorylation after 10 h at restrictive conditions (Figure 18 C). Thus, in contrast to *lrg-1*, the MAK1 levels in *cot-1* mutants are stably induced. While the phenotype of *lrg-1* was suppressed by the inhibition of PKC, no effect of the PKC inhibitors staurosporine and cercosporamide on the growth of *cot-1* was observed, suggesting that the upper part of the cell integrity signalling pathway is not affected by *cot-1* (Figure 18 D), and this may indicate an indirect effect of COT1 on MAK1. Furthermore, *cot-1(1)* grew like *wild type* on medium supplemented with 0.5 μM latrunculin A (Figure 18 E), and overexpression of $BNI1^{1-824}$ did not alter the *cot-1(1)* defects (Figure 18 F), demonstrating that COT1 is not affecting the actin organizing function of RHO1 in a manner similar to LRG1. Nevertheless, when the FH2 domain containing effector part of BNI1 (aa 1029-1817), which is predicted to regulate actin polymerisation in a dominant-active manner (Moseley et al., 2004), was overexpressed, this resulted in a synthetic death phenotype of both *lrg-1(12-20)* and *cot-1(1)* (Figure 18 G).

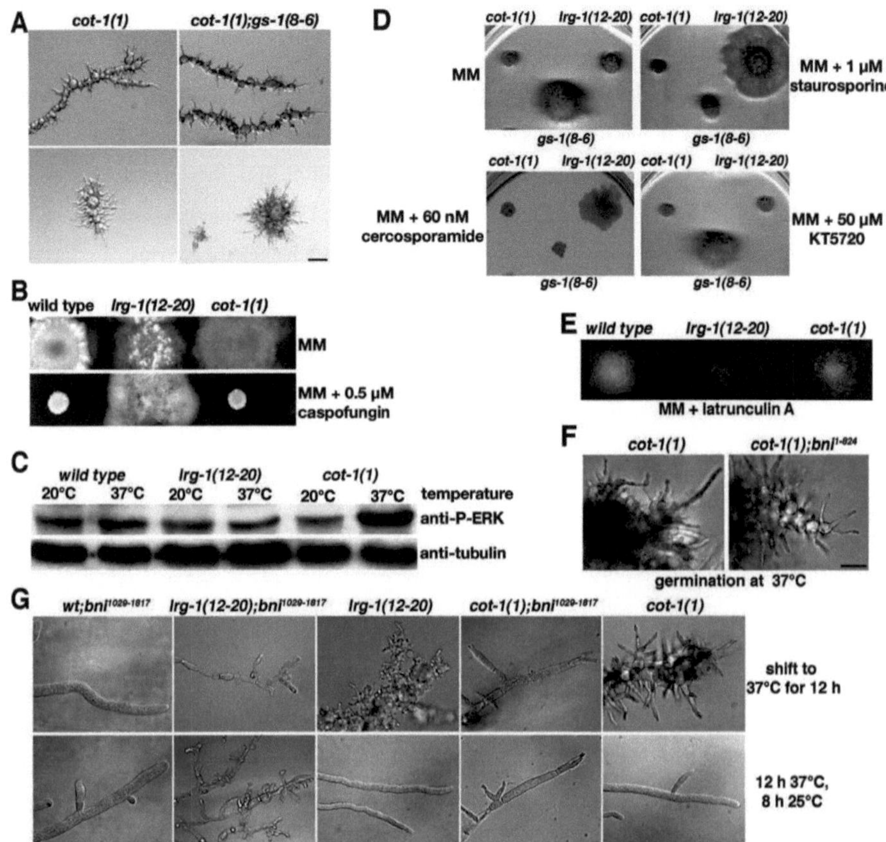

Figure 18: The Ndr kinase COT1 functions independently of RHO1 signaling.
(A) Double mutants of *cot-1(1)* and *gs-1(8-6)* did not display synthetic defects after shift to restrictive temperatures (upper panel) or when germinated under restrictive conditions (lower panel). Scale bar is 40 µm. (B) The growth behaviour of *cot-1(1)* is similiar to *wild type* on media supplemented with 0.5 μM caspofungin. (C) Total soluble protein of the indicated strains grown at 25°C or shifted to 37°C for 10 h was extracted. The blot was probed with an anti-phospho-ERK antibody, detecting increased MAK1 phosphorylation in *cot-1(1)* grown at restrictive conditions (upper panel). To confirm equal loading the blot was reprobed with anti-tubulin antibody (lower panel). (D) Staurosporine and cercosporamide partially suppress the tip extension defects of *lrg-1(12-20)*, but not those of *cot-1(1)*. (E) In contrast to *lrg-1(12-20)*, *cot-1(1)* is not hypersensitive to the addition of latrunculin A to the medium. (F) Over expression of $BNI1^{1-824}$ in *cot-1(1)* did not suppress its tip extension and hyperbranching defects. (G) Overexpression of $BNI1^{1029-1817}$ in *lrg-1(12-20)* and *cot-1(1)* was lethal at restrictive temperatures. Growth of *lrg-1(12-20);bni-1*$^{1029-1817}$ and of *cot-1(1);bni- 1*$^{1029-1817}$ was blocked within 30 min after shift to 37°C, and the strains terminated growth with tip-arrested, non-hyperbranched and vacuolated hyphae (upper panel), which were unable to resume growth at 25°C (lower panel). Scale bar in (F) and (G) is 30 µm.

With *bni-1*$^{1029\text{-}1817}$ transformed *wild type* and mutant strains were viable and generated normal growing colonies at permissive temperature. In contrast to *wild type;bni-1*$^{1029\text{-}1817}$, which was not affected by a shift to 37°C, growth of *lrg-1(12-20);bni-1*$^{1029\text{-}1817}$ and of *cot-1(1);bni-1*$^{1029\text{-}1817}$ was terminated within 30 min after the transfer with tip-arrested, non-hyperbranched and vacuolated hyphae. Further, these strains were unable to resume growth at 25°C. Thus, an effect of COT1 on the organization of the actin cytoskeleton is likely, but is distinct from the LRG1- RHO1 pathway. Taken together, the different effects observed for *lrg-1(12-20)* and *cot-1(1)* on each of the three analysed RHO1-dependent effector pathways indicate that COT1 signalling does not affect Rho1 activity.

4 Discussion

4.1 Mutations in *lrg-1* and *pod-6* affect hyphal tip elongation, septation and determination of branching in *N. crassa*

Cellular polarity is a fundamental property of every cell, and fungal hyphae are amongst the most highly polarized cells known in nature. In a visual screen several genes have been isolated, which encode components that are essential for apical tip extension and for restriction of excessive branch formation in subapical regions of the hypha (Seiler and Plamann, 2003). In this study, the functions of LRG1, COT1 and POD6 during polar tip extension were analysed in detail.

Morphological characterization of temperature sensitive and deletion mutants revealed a major defect of polar tip extension along with altered branching frequencies and increased septation when these strains were compared to *wild type*. Furthermore, all mutant strains are impaired in cell wall functions as determined by abnormal Calcofluor White distribution, by increased cell wall thickness in transmission electron microscopic sections for *pod-6(31-21)* and as reported before for *cot-1(1)* (Gorovits *et al.*, 2000) and by their growth behaviour on medium containing cell wall drugs lysing enzymes or the β1,3-glucan synthase inhibitor caspofungin.

4.2 The germinal center kinase POD6 acts together with COT1 in polar tip extension

The morphological defects of both *cot-1(1)* and *pod-6(31-21)* can be partially suppressed by various environmental stresses that decrease PKA activity and bypass the requirement for functional COT1 or POD6 (Seiler *et al.*, 2006). Both proteins can be co-immune precipitated and show a partial co-localisation. These results strongly suggest that POD6 and COT1 function in the same genetic pathway (Figure 20).

To date, the detailed hierarchical relationship between COT1 and POD6 in *N. crassa* is still an open question. The expression level of neither kinase was altered in the other mutant grown at restrictive temperature and overexpression of one kinase from the modified *N. crassa cpc-1* promoter in the other kinase mutant did not alter the mutant phenotypes. Also, no co-dependence for localization was observed (Seiler et al., 2006). This may suggest that COT1 and POD6 operate in a network rather than act in a hierarchical fashion.

Networks that regulate NDR kinase activity are also known in other systems. The *S. cerevisiae* NDR kinase Cbk1p links the regulation of cell morphology with cell-cycle progression and acts as part of the RAM network (Hergovich *et al.*, 2006). SAX1, the *C. elegans* orthologue of COT1, controls neurite outgrowth and dendritic tiling of mechanosensory neurons. SAX1 is activated by SAX2, a large conserved protein with HEAT/armadillo repeats (Gallegos and Bargmann, 2004). Interestingly, the neurite termination and tiling defects of *sax-1* null animals were significantly less severe than those of *sax-2* neurons, indicating that SAX1 may function in parallel with another kinase. This indicates that NDR kinases act in complex networks in fungi as well as in neurons. Therefore, the phenotypic complexity of filamentous fungi and cells of higher eukaryotes provides an opportunity to dissect the contribution of different COT1-interacting proteins during establishment and maintenance of cell polarity. Given the similarity between NDR and GC-III kinases in filamentous fungi and animals, *N. crassa* may serve as a useful model to separate signals required for branch formation and tip extension in neuronal cells.

Two different mechanistic explanations for the function of the COT1/POD6 complex can be proposed from the *pod-6* and *cot-1* mutants analysis. One possibility is that COT1 and POD6 act as positive regulators of tip extension, or the COT1/POD6 complex acts as inhibitory regulator of polarity establishment. The first hypothesis is supported by the morphological analysis of the mutants. Based on the mentioned observations it is conceivable that the complex functions to promote tip elongation while at the same time curbs excessive branch formation in subapical regions of the hyphal cell. This does not exclude the possibility that the involvement of COT1 and POD6 in regulation of tip elongation is more complex and is balanced with additional signal transduction pathways.

4.3 The LIM domains are required for the localization of LRG1

The domain analysis of LRG1 revealed that both regions of the protein, containing LIM and RHO-GAP domains, are essential for its cellular function. Deletion constructs of either the N-terminus ($LRG1^{1-847}$) containing the LIM domains or the C-terminus ($LRG1^{781-1279}$) containing the GAP domain are not sufficient to complement the mutant phenotype.

A single amino acid substitution of the conserved tyrosine 926 in the GAP domain in the temperature sensitive allele *lrg-1(12-20)* indicate that the GAP domain is essential for the cellular function of LRG1. A construct containing a mutation of the conserved lysine 910 to alanine that inhibits the binding of the GAP domain to the corresponding RHO protein (Li *et al.*, 1997) was not able to complement *lrg-1(12-20)* or *Δlrg-1*, indicating that binding to the corresponding RHO protein is essential for LRG1 function in tip elongation. Further, a mutation of the catalytic arginine 847 to leucine failed to complement the *lrg-1(12-20)* growth defects, indicating that the catalytic activity is essential. Taken together, these experiments show, that the binding and the catalytic activity of the LRG1 GAP domain towards the corresponding RHO protein are essential for LRG1 function. The expression of RHO proteins in *lrg-1(12-20)* resulted only for RHO1 in an initial complementation of the growth defect at restrictive temperature. As a biochemical approach, *in vitro* GAP assays identified also Rho1 as the only RHO protein in *N. crassa*, which shows enhanced GTPase activity upon stimulation by the GAP domain of LRG1. These experiments characterize LRG1 as a RHO1 specific GAP.

Several lines of evidence indicate that the N-terminal part of the protein is required for the localization of LRG1. A construct lacking the first 780 aa ($LRG1^{781-1279}$) is not capable of complementing the mutant defects, while the first 847 aa ($LRG1^{1-847}$) alone are sufficient for correct localization. The morphological defects observed in the mutant and the localization pattern of LRG1 indicate a function of LRG1 at sites of active growth at the hyphal tip and during septation. This is further supported by its growth-dependent localization at the hyphal apex and along septae and by the correlation between the growth rate and size of the GFP::LRG1 cap at the hyphal tip. A clear indication for the involvement of the three LIM domains within the N-terminus of LRG1 in the localization process is the altered localization pattern of the GFP::LRG1* protein, in which the three LIM domains have been mutated. However, despite its localization defect, GFP::LRG1* was capable of rescuing the growth defect of *Δlrg-1*, indicating additional functional patterns within the N-terminus. In this respect it is interesting to note the high sequence similarity of these domains with paxillin ($E=1e^{-28}$, while other LIM domain proteins have much weaker similarities of approximately $E=1e^{-14}$), suggesting that LRG1 might act as a cytoskeletal organizer similar to paxillin in focal adhesion complexes (Brown and Turner, 2004; Schaller, 2001). Another possibility is that the N-terminus of LRG1 interacts with microtubule-dependent motor proteins. The plus ends of the microtubule cytoskeleton are oriented towards the tip of a growing hypha (Konzack *et al.*, 2005; Schuchardt *et al.*, 2005), and conventional kinesin and dynein have been shown to drive apical and retrograde directed transport in *N. crassa*, respectively (Seiler *et al.*, 1997; Seiler *et al.*, 1999). Defective retrograde transport in the dynein heavy chain mutant resulted in increased apical localization of GFP::LRG1. Furthermore, a synthetic growth defect of *lrg-1(12-20);nkin* was observed, and in nocodazole treated cells the size of GFP::LRG1 caps was reduced at the same growth rates. The opposing transport rates of NKIN and dynein may result in a balanced distribution of LRG1 and thus allow growth that is independent of the functionality of the LIM domains. The abolishment of the microtubule and the LIM domain dependent localization mechanisms by deleting the whole N-terminus could result in the loss of function of LRG1.

A *Δlrg-1;GFP::LRG1*;ro-1* mutant strain could further contribute to clarify the influence of microtubule motors and the LIM domain function on the localisation. However, the crosses between *GFP::LRG1** and *ro-1* strains were not possible and resulted in a developmental arrest at the stage of fertilized perithecia.

4.4 LRG1 regulates the activity of several Rho1 effector pathways

Rho1 is a key regulator of hyphal growth and polarity and has been described as integral part of the β1,3-glucan synthase (GS) complex in yeast and filamentous fungi (Arellano *et al.*, 1996; Beauvais *et al.*, 2001; Mazur and Baginsky, 1996; Qadota *et al.*, 1996). Furthermore, Pkc1p and the formin proteins Bni1p and Bnr1p are known targets of Rho1p in *S. cerevisiae*. Pkc1p activates several signal transduction pathways including the cell wall integrity MAP kinase Mak1p. Formin proteins are activated by binding to Rho GTPases by their GTPase binding domain (GBD) and subsequently act as nucleus for polymerisation of linear actin cables (summarized in Levin, 2005).
The *in vivo* results presented in this study indicate several misregulated RHO1 effector pathways in strains lacking functional LRG1. The observed hyposensitivity to the β1,3-glucan synthase inhibitor caspofungin and the genetic interaction of *lrg-1(12-20)* with *gs-1(8-6)* are consistent with increased β1,3-glucan synthase activity in *lrg-1(12-20)* and indicate that GS activity is regulated via the GAP LRG1.
Furthermore, the latrunculin A sensitivity of *lrg-1(12-20)* and the partial suppression of the *lrg-1(12-20)* growth defects by the dominant-negatively acting $BNI1^{1-824}$ containing the GBD indicate that RHO1-BNI1-actin signalling is affected in *lrg-1(12-20)*. $BNI1^{1029-1817}$ contains the FH2 effector domain and is suggested to act in a dominant active fashion in actin polymerization. Expression of $BNI1^{1029-1817}$ in *lrg-1(12-20)* resulted in synthetic defects. This suggests, that the formin mediated actin polymerisation is disturbed in *lrg-1(12-20)*.
The suppression of *lrg-1(12-20)* by the PKC specific inhibitors staurosporine and cercosporamide indicate hyperactivity of PKC in *lrg-1(12-20)*. Interesting is that the activity of the downstream MAK1 pathway is not induced in *Δlrg-1* or *lrg-1(12-20)*, suggesting that the MAP kinase cascade is not regulated by LRG1. However, the MAK1 activation pattern in heat stressed *lrg-1(12-20)* is different from wild type, which may support a regulatory function of LRG1 on the cell integrity pathway. If this is the case, additional feed back mechanisms are postulated that regulate the activation of the PKC – MAK1 part of the cell integrity pathway in *N. crassa*. In contrast to *mak-1*, *pkc* is essential in *N. crassa*, indicating that PKC influences additional important effector pathways other than MAK1. Therefore, the observed differences between PKC and MAK1 activity would argue for a regulation of specific PKC effectors by LRG1 (Figure 19), and suggest that the signal is

not transmitted to the MAK1 MAP kinase pathway. Based on the presented data, I propose that LRG1 regulates RHO1-PKC signalling in *Neurospora crassa,* although this activation may not be transmitted towards the MAK1 MAP kinase cascade.

One important question concerning the investigated Rho1 signal transduction pathways is, which of these RHO1 effector pathways is primarily responsible for the block in tip extension and the subapical hyperbranching observed in *lrg-1(12-20)*. Overexpression as well as deletion of components of PKC/MAK1 MAP kinase pathway results in morphological changes different from those described here for *lrg-1* mutants (Stephan Seiler, personal communication), and thus, impairment of the PKC/MAK1 pathway by LRG1 deficiency is probably not the primary reason for the block in tip extension and the induction of new hyphal tips along the entire cell. Nevertheless, the unregulated deposition of excessive chitin within the cell wall of *lrg-1(12-20)* may result from changed activity of the PKC pathway. As demonstrated in budding yeast, cell wall stress can result in an up to tenfold increase in the chitin content of the cell wall by stress-induced mobilization of chitin synthase III (Chs3p) from chitosome vesicles to the plasma membrane (Valdivia and Schekman, 2003). This stress signal is transmitted via Rho1p and involves components of the cell integrity pathway including Pkc1p, but is independent of Mak1p signalling (Valdivia and Schekman, 2003; summarized in Levin, 2005). Therefore the observed hyperactivity of PKC without affecting MAK1 signalling is consistent with the abnormal chitin deposition in *lrg-1(12-20)*.

Loss of the GAP activity of LRG1 is predicted to results in the activation of RHO1 and therefore increased glucan synthase activity. As predicted, increased glucan deposition in the cell walls of *lrg-1(12-20)* and *gs-1(8-6)* was detected by the hyposensitivity to the glucan synthase inhibitor caspofungin. Thus, the common hyperbranching phenotype observed in *lrg-1(12-20)* and *gs-1(8-6)* suggests that increased glucan synthase activity can result in the induction of branch formation.

Data in *S. cerevisiae* argue for a connection between the actin cytoskeleton and cell wall synthesis. A detailed localization study in *S. cerevisiae* revealed that GS colocalizes with cortical actin patches and moves on the cell surface in a manner dependent on actin patch mobility (Utsugi *et al.*, 2002). Actin patches are associated with the invaginated plasma membrane (Mulholland *et al.*, 1994; Rodal *et al.*, 2005), consistent with the role of actin at the internalization step of endocytosis (Kubler and Riezman, 1993). Endocytosis regulates cell wall assembly by controlling the levels of plasma membrane-associated, cell wall synthesizing enzymes such as chitin synthase III (Chs3p) (Valdivia *et al.*, 2002; Ziman *et al.*, 1996) and β1,3-glucan synthase (Fks1p) (Engqvist-Goldstein and Drubin, 2003; Utsugi *et al.*, 2002). Furthermore cell wall synthetic enzymes are transported and distributed by the actin cytoskeleton in *S. cerevisiae* (Utsugi *et al.*, 2002).

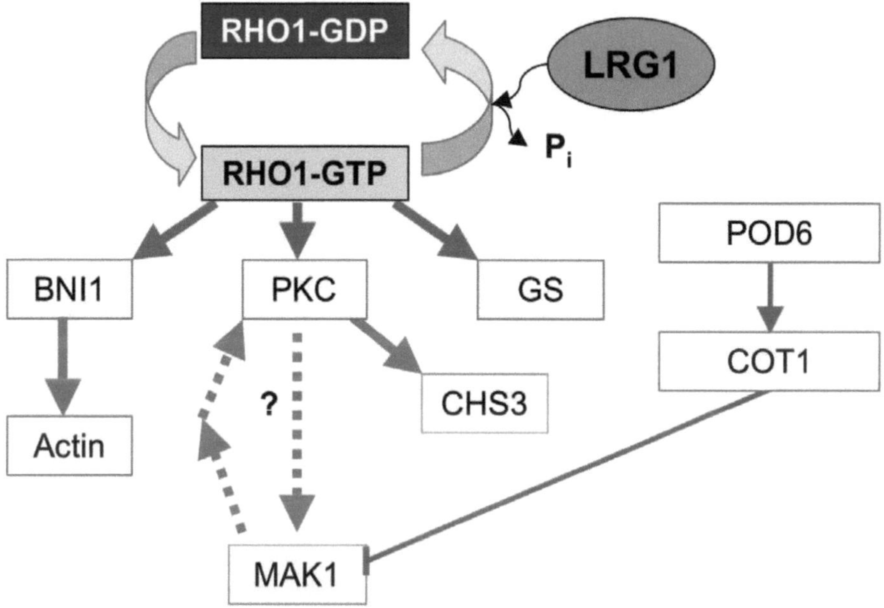

Figure 19 (former page): The model of potential connections of LRG1, COT1 and POD6 concerning polar tip growth in *N. crassa*.
POD6 and COT1 act together in one pathway. POD6 represents possibly an upstream kinase of COT1 acting on the phosphorylation site in the hydrophobic motive. The COT1 dependent signal transduction represses MAK1 phosphorylation by an unknown mechanism that may involve PAK kinases, but is PKC independent. The glucan synthase (GS), the formin BNI1 and PKC are targets of RHO1. The potential effectors of the PKC may include the chitin synthase 3 (CHS3). LRG1 acts on Rho1, thereby inactivating the Rho GTPase. All tested effectors, BNI1, PKC and GS, are influenced by LRG1. The enhanced PKC activity does not result (as suggested from data in *S. cerevisiae*) in MAK1 phosphorylation, indicating additional or different regulations that influence MAK1 activity.

Interestingly, the *lrg-1(12-20);gs-1(8-6)* double mutant loses cell polarity when hyphae are shifted to restrictive conditions and is unable to establish polarity when conidia are germinated at high temperature. This phenotype is reminiscent of conditional mutants of key regulators of the actin cytoskeleton such as *bem-1* and *cdc-24* (Seiler and Plamann, 2003) and suggests a connection between cell wall function and the actin cytoskeleton (Garcia *et al.*, 2006a; Katayama *et al.*, 1999). Moreover, a microscopic analysis of *lrg-1(12-20);bni-1*$^{1\text{-}824}$ revealed a phenotype related to *gs-1(8-6)* (Figure 17 E). This GTPase binding domain (GBD) containing part of BNI1 can act through quenching Rho1 activity by competition with other target molecules for RHO1 binding. Formin proteins form autoinhibitory intramolecular interactions, that can be released by binding of the formin to the Rho protein (Alberts, 2001; Palazzo *et al.*, 2001). When the N-terminal part is overexpressed, heterodimers may form and inhibit the endogenous formin proteins irreversible. This suggests that increased GS1 and BNI1 activity are two major causes for the *lrg-1* defects.

4.5 Comparison of LRG1 to its homologues in yeasts

A homologue of LRG1 is found in *S. cerevisiae*, but conflicting evidence about the Rho proteins and the Rho effector pathways that are regulated by Lrg1p was reported. *In vitro* and yeast two hybrid analyses identified Lrg1p as a GAP for Rho1p, Rho2p or Cdc42p or for combinations of these GTPases (Fitch *et al.*, 2004; Lorberg *et al.*, 2001; Roumanie *et al.*, 2001; Watanabe *et al.*, 2001). However, most genetic data link Lrg1p with Rho1p functions (Fitch *et al.*, 2004; Lorberg *et al.*, 2001; Stewart *et al.*, 2007; Varelas *et al.*, 2006; Watanabe *et al.*, 2001), suggesting that Lrg1p is a Rho1p specific GAP in yeast. The effector pathways that are regulated by Lrg1p are discussed in a controversial manner. Several studies report enhanced PKC/MAP kinase activity (Lorberg *et al.*, 2001; Stewart *et al.*, 2007; Varelas *et al.*, 2006) in contrast to a report describing no effect on the cell wall integrity pathway (Watanabe *et al.*, 2001). In addition, increased glucan synthase activity was reported for *Δlrg-1* (Fitch *et al.*, 2004; Lorberg *et al.*, 2001; Watanabe *et al.*, 2001), and no effect of Lrg1p on the formin to actin branch of Rho1p signalling was reported in any of these studies. In contrast, Nakano and coworkers (2001) reported delocalized actin patches in a mutant defective in *rga-1*, the fission yeast homologue of *lrg-1*. In conclusion, the results from both yeasts and the data presented in this study suggest that LRG1 acts as a GAP specific for RHO1 and regulates several RHO1-dependent effector pathways in *N. crassa* and possibly also in other fungi.

4.6 The COT1/POD6 complex and LRG1 act in parallel morphogenetic pathways

The characterization of *pod-6(31-21)*, *Δpod-6*, *lrg-1(12-20)* and *Δlrg-1* revealed that deletion of *pod-6* confers the same and deletion of *lrg-1* confers similar defects concerning branching, septation, tip growth and morphology as those previously characterized in *cot-1(1)* (Collinge *et al.*, 1978; Collinge and Trinci, 1974; Yarden *et al.*, 1992). In addition a *pod-6(31-21);cot-1(1)* double mutant exhibits phenotypic characteristics identical to the parental strains (Figure 8 B) in contrast to *lrg-1(12-20);cot-1(1)* and *lrg-1(12-20);pod-6(31-21)* double mutant strains that show synthetic defects (Figure 17 A). Furthermore, *pod-6(31-21)* and *cot-1(1)* share the putative phosphatase *gul-1* and components of cytoplasmic dynein as common extragenic suppressors (Seiler *et al.*, 2006), while *lrg-1(12-20)* share the components of the dynein complex, but not *gul-1*, as suppressors. Taken together, these data indicate that *cot-1* and *pod-6* may act in a common pathway in parallel to *lrg-1*, and that both are required for polar tip extension, branching and septation in *N. crassa*.

COT1 and POD6 localise in a punctate pattern throughout the hypha and are enriched at tips and at septae (Seiler *et al.*, 2006). In *S. pombe* the NDR kinase Orb6p is also enriched at sites of cell growth (Hirata *et al.*, 2002; Verde *et al.*, 1998; Wiley *et al.*, 2003). In immune localisation experiments, LRG1 was detected on septae and similar to COT1 and POD6 enriched at hyphal tips. However the localisation at the cortex of hyphal tips was only observed with the GFP tagged LRG1 in living cells. Therefore, it would be interesting to find out, whether COT1 and POD6 concentrations increase on the tip cortex by using GFP tagged proteins in living cells. Due to the similar immune localisation, further studies should also address the question, whether LRG1 colocalises with COT1 and POD6.

Beside the partial colocalisation, COT1 and POD6 are misslocalised in a dynein heavy chain mutant to the tip and in a kinesin mutant to the septae (Seiler *et al.*, 2006). Thus, equal distribution of the NDR and GCK kinases COT1 and POD6 is dependent on microtubule dependent motor proteins. A similar dependence of localisation pattern was seen for the GFP::LRG1 protein in the dynein mutant background (Figure 17 C) and the synthetic growth defects with kinesin indicate a motor protein dependent distribution also for LRG1.

The morphological similarities between *lrg-1(12-20), pod-6(31-21)* and *cot-1(1)* at restrictive temperature and the common dependence of the localisation of POD6, COT1 and LRG1 on microtubule driven transport indicated a connection between COT1 and LRG1 signalling pathways. Genetic data argue for a connection between Ndr kinases and the actin cytoskeleton and link *C. elegans sax-1* and *D. melanogaster trc* with Rho protein function (Emoto *et al.*, 2004; Zallen *et al.*, 2000).

Genetic data in *S. cerevisiae* suggest that the COT1 homologue Cbk1p may negatively regulate the small GTPase Rho1p, which in turn activates the cell wall integrity pathway that is most similar to the *N. crassa* MAK1 pathway (Emoto *et al.*, 2004; Jorgensen *et al.*, 2002; Schneper *et al.*, 2004; Versele and Thevelein, 2001; Zallen *et al.*, 2000). A physical interaction has also been shown to exist between the Ndr kinase ORB6 and a RHO-GTPase activating protein in fission yeast (Das *et al.*, 2007). The observation that in *cbk1* mutants actin localizes at many cortical patches independent of the growing tip (Weiss *et al.*, 2002) further indicates a possible influence of NDR kinases on Rho protein activity.

However, no influence of COT1 on RHO1 effectors was observed in this work. Double mutants of *cot-1(1)* with *gs-1(8-6)* showed no synthetic effect and *cot-1(1)* was as sensitive as wild type to caspofungin, indicating that the glucan content in *cot-1(1)* was not altered. The sensitivity toward the actin depolymerising drug latrunculin A was not altered in *cot-1(1)*. Also the expression of the N-terminus of BNI1 ($BNI1^{1\text{-}824}$) in *cot-1(1)* had, in contrast to the expression in *lrg-1(12-20)*, no effect. This indicates that the formin dependent actin polymerisation is not disturbed in *cot-1(1)*.

The PKC inhibitors cercosporamide and staurosporine compensated only the phenotype of *lrg-1(12-20)*, but not of *cot-1(1)*, indicating that PKC is only hyperactive in *lrg-1(12-20)*, but not in *cot-1(1)*. Only the MAK1 kinase was misregulated in *cot-1(1)*, and this in a different way from that in *lrg-1(12-20)*. Recently, the binding of the NDR kinase Stk38 to Mekk was reported in mammalian cells (Enomoto *et al*., 2007), indicating a possible regulation of MAK by NDR kinases that is independent of PKC activation. This provides evidence that the PKC is not activated by RHO1 in *cot-1(1)*.

Taken together, these data indicate that the common mutant defects and the common dependence for the localisation of COT1/POD6 and LRG1 on microtubule motors are at least in part based on different signal transduction pathways. However, a common target of both signalling routes is the cell wall, suggesting that a balanced composition of cell wall components is critical for directed growth in *N. crassa*.

4.7 The influence of RHO cycling on activity

Rho proteins are well known regulators of a variety of fundamental cellular processes. However, to date only RHO1, CDC42 and RAC are well characterised in several model systems. One main goal for the future would be to understand the function and regulation of all Rho proteins present in *N. crassa*. The established purification procedure for the sensitive *N. crassa* Rho proteins and the *in vitro* assays for GAP and GEF activity are described in this study. They can be used for *in vitro* characterisation of the biochemical activity of all annotated GAP and GEF proteins towards the six Rho GTPases in *N. crassa*.

Furthermore, the generation of deletion mutants (Colot *et al*., 2006; Dunlap *et al*., 2007) of all RHO proteins, GAPs and GEFs and at least some effectors will contribute to find phenotypic similarities and genetic interactions between regulators and RHO proteins and alleles that show cross complementation. This will help to further understand the RHO regulation network in *N. crassa*.

Remarkably, overexpression of *wild type* and of both dominant *rho-1* versions partially compensates the *lrg-1(12-20)* growth defects. Similar results are known for the related GTPase CDC42. Stable overexpression of the constitutively active $Cdc42^{G12V}$ mutant resulted in inhibition of cell proliferation in mammalian cells, which is a dominant negative phenotype (Fidyk *et al.*, 2006; Vanni *et al.*, 2005). Also in yeast cycling of Cdc42p was required for efficient cell fusion in the mating process (Barale *et al.*, 2006). These studies concluded that the complete GTPase cycle is required for full G-protein activity. The fact that *lrg-1(12-20)* cells, which are assumed to be deficient in Rho-GAP activity, are defective in polarized growth also suggest that cycling of RHO1, rather than its GTP-bound form alone is important for its function. Thus the suppression of the *lrg-1(12-20)* phenotype by both dominant alleles of RHO1 further suggests that polar tip extension and the regulation of branch formation involves cycling of *N. crassa* RHO1 to gain full activation, rather than functioning as simple on/off switch.

In this context a computational model was developed by Goryachev and Pokhilko (2006). It predicts that the activity of CDC42 at physiological concentrations in the absence of GAP proteins depends mainly on the GEF concentration. Expressing unphysiologically high amounts of CDC42 at constant GEF concentration reduces its relative activity, which is mainly due to titration of the GEF and dilution of the active by the inactive form of CDC42. This will decrease the relative activity of CDC42 in active cellular zones. The temporary rescue of the *lrg-1(12-20)* defects by expressing high amounts of both, constitutively active and negative alleles of RHO1 is consistent with this prediction.

One open question is, why the cycling is required for proper activity of the effector pathways. The cycling of the nucleotide state of the Rho protein is important for the change of the affinity of Rho proteins towards GDI (guanidine dissociation inhibitor) proteins. These proteins serve as cytoplasmic storage for RHO GTPases and are possibly involved in transport processes to sites of active protein pools (reviewed in Dovas and Couchman, 2005; Olofsson, 1999). These authors speculate that GDI complexes may deliver RHO proteins to the membrane compartments, where they are activated by GEFs.

These speculations can be integrated in a model (Figure 20) that requires shuttling between the GTP and GDP bound form of RHO proteins to allow the GDP bound form to be transported by GDI through the cytoplasm to sites of active RHO protein. The displacement from the GDI is suggested to deliver the Rho proteins directly to GEF containing sites and may be regulated by the lipid composition at these sites (Dovas and Couchman, 2005). This would in part explain the observed requirement of cycling RHO proteins for their full activity at sites of active growth.

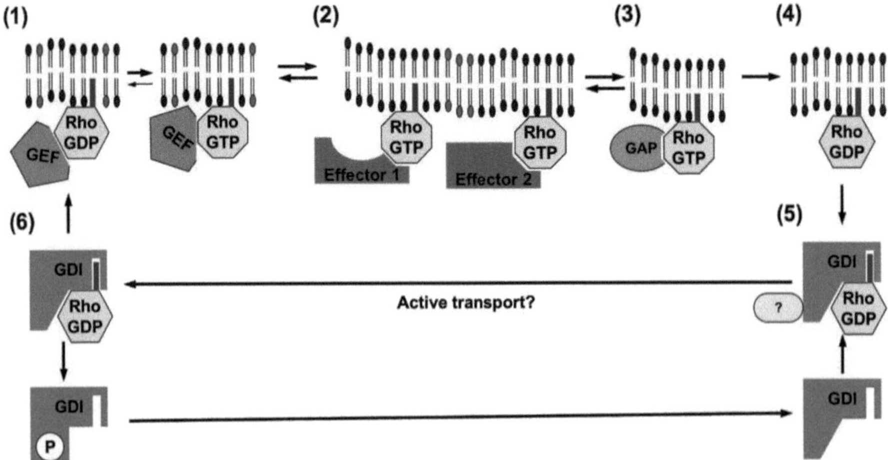

Figure 20 (former page): Shuttling model of GTPase cycle.
(1) At sites of high Rho activity, the lipid composition (lipids special for these sites in red) and possibly unknown interaction partners recruit the cytoplasmic GDI-Rho complexes, which leads to activation of the Rho proteins due to the GEF regulators. (2) The active Rho protein interacts with several effectors. (3) At flanking sites of inactivation, the GAP activity is increased by several mechanisms, including for example phosphorylation of GAPs, altered lipid composition of the membrane and local increased GAP concentrations. (4) The GDP bound form of the Rho protein has higher affinity to the GDI than the GTP bound form, which is additionally increased by altered lipid composition in the membrane compared to active Rho compartments. (5) The GDI with inactive Rho may be in complex with unknown factors (yellow), which contribute to the recognition of target compartments in the membrane. It is possible, that active transport is involved. (6) The GDI delivers the Rho protein to special sites at the cell cortex on the membrane. This may be enhanced by phoshorylation of the GDI.
These regulations enable the cell to specially increase the active pool of Rho protein in a very restricted area and therefore essentially require the GTP-GDP transition of the Rho protein for full activation only at the active site. Further, the locally available active Rho protein can be fast delivered through the cytoplasm.

In this context, a long-term goal would be to differentiate between the active part and the whole pool of Rho protein within a living cell. The visualisation of the activity of Rho1 could further address the question whether the *cot-1(1)* and *pod-6(31-21)* mutants influence Rho1 activity and how the local activity is altered in *lrg-1(12-20)* mutants.

The overexpression of LRG1 or its GAP domain had no obvious effect on growth and morphology (data not shown), and *in vitro* GTPase activity of RHO1 was specifically induced by LRG1$^{650\text{-}1035}$. A catalytic inactive mutant R847L of this domain fused to GFP may therefore specifically recognise GTP bound RHO1 and would be a helpful tool for further *in vivo* experiments. However, if this construct will specifically recognise active RHO1 and not localise at different sites of the cell, has to be shown. Alternatively, an RHO1 binding domain of an effector may be used in such an approach.

5 References

Adamo, J. E., Rossi, G. and Brennwald, P. (1999). The Rho GTPase Rho3 has a direct role in exocytosis that is distinct from its role in actin polarity. Mol Biol Cell 10, 4121-4133.

Adler, P. N. (2002). Planar signaling and morphogenesis in *Drosophila*. Dev Cell 2, 525-535.

Alberts, A. S. (2001). Identification of a carboxyl-terminal diaphanous-related formin homology protein autoregulatory domain. J Biol Chem 276, 2824-2830.

Arellano, M., Duran, A. and Perez, P. (1996). Rho 1 GTPase activates the (1-3)beta-D-glucan synthase and is involved in *Schizosaccharomyces pombe* morphogenesis. EMBO J 15, 4584-4591.

Ausübel, F., Brent, R., Kingston, R., Moore, D., Seidman, J., Smith, J., Struhl, K., Albright, L., Coen, D. and Varki, A. (eds.) (1997). Current Protocols in Molecular Biology. John Wiley & Sons, New York.

Axelrod, J. D. and McNeill, H. (2002). Coupling planar cell polarity signaling to morphogenesis. ScientificWorldJournal 2, 434-454.

Bach, I. (2000). The LIM domain: regulation by association. Mech Dev 91, 5-17.

Bähler, J. and Peter, M. (2000). Cell Polarity in yeast. Cell Polarity 21-77.

Barale, S., McCusker, D. and Arkowitz, R. A. (2006). Cdc42p GDP/GTP cycling is necessary for efficient cell fusion during yeast mating. Mol Biol Cell 17, 2824-2838.

Beauvais, A., Bruneau, J. M., Mol, P. C., Buitrago, M. J., Legrand, R. and Latge, J. P. (2001). Glucan synthase complex of *Aspergillus fumigatus*. J Bacteriol 183, 2273-2279.

Bechet, J., Greenson, M. and Wiame, J. M. (1970). Mutations affecting the repressibility of arginine biosynthetic enzymes in *Saccharomyces cerevisiae*. Eur J Biochem 12, 31-39.

Bidlingmaier, S., Weiss, E. L., Seidel, C., Drubin, D. G. and Snyder, M. (2001). The Cbk1p pathway is important for polarized cell growth and cell separation in *Saccharomyces cerevisiae*. Mol Cell Biol 21, 2449-2462.

Bokoch, G. M. (2003). Biology of the p21-activated kinases. Annu Rev Biochem 72, 743-781.

Bombarda, E., Cherradi, H., Morellet, N., Roques, B. P. and Mely, Y. (2002). Zn2+ binding properties of single-point mutants of the C-terminal zinc finger of the HIV-1 nucleocapsid protein: evidence of a critical role of cysteine 49 in Zn2+ dissociation. Biochemistry 41, 4312-4320.

Borkovich, K. A., Alex, L. A., Yarden, O., Freitag, M., Turner, G. E., Read, N. D., Seiler, S., Bell-Pedersen, D., Paietta, J., Plesofsky, N. *et al.* (2004). Lessons from the genome sequence of *Neurospora crassa:* tracing the path from genomic blueprint to multicellular organism. Microbiol Mol Biol Rev 68, 1-108.

Bowman, S. M. and Free, S. J. (2006). The structure and synthesis of the fungal cell wall. Bioessays 28, 799-808.

Boyce, K. J. and Andrianopoulos, A. (2006). Morphogenesis: Control of Cell Types and Shape. In: The Mycota I: Growth, Differentiation and Sexuality, 3-20. K. Esser, U. Kües and R. Fischer (eds.), Springer-Verlag Berlin Heidelberg, Berlin.

Boyce, K. J., Hynes, M. J. and Andrianopoulos, A. (2003). Control of morphogenesis and actin localization by the *Penicillium marneffei* RAC homolog. J Cell Sci 116, 1249-1260.

Boyce, K. J., Hynes, M. J. and Andrianopoulos, A. (2001). The CDC42 homolog of the dimorphic fungus *Penicillium marneffei* is required for correct cell polarization during growth but not development. J Bacteriol 183, 3447-3457.

Boyce, K. J., Hynes, M. J. and Andrianopoulos, A. (2005). The Ras and Rho GTPases genetically interact to co-ordinately regulate cell polarity during development in *Penicillium marneffei*. Mol Microbiol 55, 1487-1501.

Brachmann, C. B., Davies, A., Cost, G. J., Caputo, E., Li, J., Hieter, P. and Boeke, J. D. (1998).

Designer deletion strains derived from *Saccharomyces cerevisiae* S288C: a useful set of strains and plasmids for PCR-mediated gene disruption and other applications. Yeast 14, 115-132.

Brennwald, P. and Rossi, G. (2007). Spatial regulation of exocytosis and cell polarity: yeast as a model for animal cells. FEBS Lett 581, 2119-2124.

Brown, M. C. and Turner, C. E. (2004). Paxillin: adapting to change. Physiol Rev 84, 1315-1339.

Bruno, K. S., Tinsley, J. H., Minke, P. F. and Plamann, M. (1996). Genetic interactions among cytoplasmic dynein, dynactin, and nuclear distribution mutants of *Neurospora crassa*. Proc Natl Acad Sci U S A 93, 4775-4780.

Chen, C. and Dickman, M. B. (2004). Dominant active Rac and dominant negative Rac revert the dominant active Ras phenotype in *Colletotrichum trifolii* by distinct signalling pathways. Mol Microbiol 51, 1493-1507.

Chilton, J. K. (2006). Molecular mechanisms of axon guidance. Dev Biol 292, 13-24.

Collinge, A. J., Fletcher, M. H. and Trinci, A. P. J. (1978). Physiological and cytology of septation and branching in a temperature-sensitive colonial mutant (*cot-1*) of *Neurospora crassa*. Trans Br Mycol Soc 71, 107-120.

Collinge, A. J. and Trinci, A. P. (1974). Hyphal tips of wild-type and spreading colonial mutants of *Neurospora crassa*. Arch Microbiol 99, 353-368.

Colot, H. V., Park, G., Turner, G. E., Ringelberg, C., Crew, C. M., Litvinkova, L., Weiss, R. L., Borkovich, K. A. and Dunlap, J. C. (2006). A high-throughput gene knockout procedure for *Neurospora* reveals functions for multiple transcription factors. Proc Natl Acad Sci U S A 103, 10352-10357.

Corpet, F. (1988). Multiple sequence alignment with hierarchical clustering. Nucleic Acids Res 16, 10881-10890.

Cote, J. F. and Vuori, K. (2007). GEF what? Dock180 and related proteins help Rac to polarize cells in new ways. Trends Cell Biol 17, 383-393.

Dan, I., Ong, S. E., Watanabe, N. M., Blagoev, B., Nielsen, M. M., Kajikawa, E., Kristiansen, T. Z., Mann, M. and Pandey, A. (2002). Cloning of MASK, a novel member of the mammalian germinal center kinase III subfamily, with apoptosis-inducing properties. J Biol Chem 277, 5929-5939.

Dan, I., Watanabe, N. M. and Kusumi, A. (2001). The Ste20 group kinases as regulators of MAP kinase cascades. Trends Cell Biol 11, 220-230.

Das, M., Wiley, D. J., Medina, S., Vincent, H. A., Larrea, M., Oriolo, A. and Verde, F. (2007).

Regulation of cell diameter, For3p localization, and cell symmetry by fission yeast Rho-GAP Rga4p. Mol Biol Cell 18, 2090-2101.

Davis, R. D. and DeSerres, F. J. (1970). Genetic and microbiological research techniques for *Neurospora crassa.* Methods Enzymol 17, 79-143.

Denning, D. W. (2003). Echinocandin antifungal drugs. Lancet 362, 1142-1151.

Dighton, J. (2007). Nutrient Cycling by Saprotrophic Fungi in Terrestrial Habitats. In: The Mycota IV: Environmental and Microbial Relationships, 287-300. K. Esser, C. P. Kubicek and I. S. Druzhinina (eds.), Springer-Verlag Berlin Heidelberg, Berlin.

Divon, H. H. and Fluhr, R. (2007). Nutrition acquisition strategies during fungal infection of plants. FEMS Microbiol Lett 266, 65-74.

Doignon, F., Weinachter, C., Roumanie, O. and Crouzet, M. (1999). The yeast Rgd1p is a GTPase activating protein of the Rho3 and Rho4 proteins. FEBS Lett 459, 458-462.

Dovas, A. and Couchman, J. R. (2005). RhoGDI: multiple functions in the regulation of Rho family GTPase activities. Biochem J 390, 1-9.

Drubin, D. G. and Nelson, W. J. (1996). Origins of cell polarity. Cell 84, 335-344.

Du, L. L. and Novick, P. (2002). Pag1p, a novel protein associated with protein kinase Cbk1p, is required for cell morphogenesis and proliferation in *Saccharomyces cerevisiae*. Mol Biol Cell 13, 503-514.

Dunlap, J. C., Borkovich, K. A., Henn, M. R., Turner, G. E., Sachs, M. S., Glass, N. L., McCluskey, K., Plamann, M., Galagan, J. E., Birren, B. W. *et al.* (2007). Enabling a community to dissect an organism: overview of the *Neurospora* functional genomics project. Adv Genet 57, 49-96.

Emoto, K., He, Y., Ye, B., Grueber, W. B., Adler, P. N., Jan, L. Y. and Jan, Y. N. (2004). Control of dendritic branching and tiling by the Tricornered-kinase/Furry signaling pathway in *Drosophila* sensory neurons. Cell 119, 245-256.

Emoto, K., Parrish, J. Z., Jan, L. Y. and Jan, Y. N. (2006). The tumour suppressor Hippo acts with the NDR kinases in dendritic tiling and maintenance. Nature 443, 210-213.

Engqvist-Goldstein, A. E. and Drubin, D. G. (2003). Actin assembly and endocytosis: from yeast to mammals. Annu Rev Cell Dev Biol 19, 287-332.

Enomoto, A., Kido, N., Ito, M., Morita, A., Matsumoto, Y., Takamatsu, N., Hosoi, Y. and Miyagawa, K. (2007). Negative regulation of MEKK1/2 signaling by Serine-Threonine kinase 38 (STK38). Oncogene, 1476-5594.

Eshel, D., Urrestarazu, L. A., Vissers, S., Jauniaux, J. C., van Vliet-Reedijk, J. C., Planta, R. J. and Gibbons, I. R. (1993). Cytoplasmic dynein is required for normal nuclear segregation in yeast. Proc Natl Acad Sci U S A 90, 11172-11176.

Etienne-Manneville, S. and Hall, A. (2002). Rho GTPases in cell biology. Nature 420, 629-635.

Evangelista, M., Zigmond, S. and Boone, C. (2003). Formins: signaling effectors for assembly and polarization of actin filaments. J Cell Sci 116, 2603-2611.

Fidyk, N., Wang, J. B. and Cerione, R. A. (2006). Influencing cellular transformation by modulating the rates of GTP hydrolysis by Cdc42. Biochemistry 45, 7750-7762.

Fitch, P. G., Gammie, A. E., Lee, D. J., de Candal, V. B. and Rose, M. D. (2004). Lrg1p is a Rho1 GTPase-activating protein required for efficient cell fusion in yeast. Genetics 168, 733-746.

Gadd, G. M. (2007). Fungi and Industrial Pollutants. In: The Mycota IV: Environmental and Microbial Relationships, 69-84. K. Esser, C. P. Kubicek and I. S. Druzhinina (eds.), Springer-Verlag Berlin Heidelberg, Berlin.

Gallegos, M. E. and Bargmann, C. I. (2004). Mechanosensory neurite termination and tiling depend on SAX-2 and the SAX-1 kinase. Neuron 44, 239-249.

Gamauf, C., Metz, B. and Seiboth, B. (2007). Degradation of Plant Cell Wall Polymers by Fungi. In: The Mycota IV: Environmental and Microbial Relationships, 325-340. K. Esser, C. P. Kubicek and I. S. Druzhinina (eds.), Springer-Verlag Berlin Heidelberg, Berlin.

Garcia, P., Tajadura, V., Garcia, I. and Sanchez, Y. (2006a). Rgf1p is a specific Rho1-GEF that coordinates cell polarization with cell wall biogenesis in fission yeast. Mol Biol Cell 17, 1620-1631.

Garcia, P., Tajadura, V., Garcia, I. and Sanchez, Y. (2006b). Role of Rho GTPases and Rho-GEFs in the regulation of cell shape and integrity in fission yeast. Yeast 23, 1031-1043.

Garcia-Rodriguez, L. J., Gay, A. C. and Pon, L. A. (2006). Organelle Inheritance in Yaests and other Fungi. In: The Mycota I: Growth, Differentiation and Sexuality, 21-36. K. Esser, U. Kües and R. Fischer (eds.), Springer-Verlag Berlin Heidelberg, Berlin.

Geng, W., He, B., Wang, M. and Adler, P. N. (2000). The *tricornered* gene, which is required for the integrity of epidermal cell extensions, encodes the *Drosophila* nuclear DBF2-related kinase. Genetics 156, 1817-1828.

Gibbs, J. B. (1995). Determination of guanine nucleotides bound to Ras in mammalian cells. Methods Enzymol 255, 118-125.

Gibbs, J. B., Schaber, M. D., Allard, W. J., Sigal, I. S. and Scolnick, E. M. (1988). Purification of ras GTPase activating protein from bovine brain. Proc Natl Acad Sci U S A 85, 5026-5030.

Gloer, J. B. (2007). Applications of Fungal Ecology in the Search for New Bioactive Natural Products. In: The Mycota IV: Environmental and Microbial Relationships, 257-286. K. Esser, C. P. Kubicek and I. S. Druzhinina (eds.), Springer-Verlag Berlin Heidelberg, Berlin.

Goldstein, A. L. and McCusker, J. H. (1999). Three new dominant drug resistance cassettes for gene disruption in *Saccharomyces cerevisiae*. Yeast 15, 1541-1553.

Gorovits, R., Propheta, O., Kolot, M., Dombradi, V. and Yarden, O. (1999). A mutation within the catalytic domain of COT1 kinase confers changes in the presence of two COT1 isoforms and in Ser/Thr protein kinase and phosphatase activities in *Neurospora crassa*. Fungal Genet Biol 27, 264-274.

Gorovits, R., Sjollema, K. A., Sietsma, J. H. and Yarden, O. (2000). Cellular distribution of COT1 kinase in *Neurospora crassa*. Fungal Genet Biol 30, 63-70.

Gorovits, R. and Yarden, O. (2003). Environmental suppression of *Neurospora crassa cot-1* hyperbranching: a link between COT1 kinase and stress sensing. Eukaryot Cell 2, 699-707.

Goryachev, A. B. and Pokhilko, A. V. (2006). Computational model explains high activity and rapid cycling of Rho GTPases within protein complexes. PLoS Comput Biol 2, e172.

Guest, G. M., Lin, X. and Momany, M. (2004). *Aspergillus nidulans* RhoA is involved in polar growth, branching, and cell wall synthesis. Fungal Genet Biol 41, 13-22.

Gulli, M. P. and Peter, M. (2001). Temporal and spatial regulation of Rho-type guanine-nucleotide exchange factors: the yeast perspective. Genes Dev 15, 365-379.

Hakoshima, T., Shimizu, T. and Maesaki, R. (2003). Structural basis of the Rho GTPase signaling. J Biochem 134, 327-331.

Hall, A. (2005). Rho GTPases and the control of cell behaviour. Biochem Soc Trans 33, 891-895.

Harris, S. D. (2006). Cell polarity in filamentous fungi: shaping the mold. Int Rev Cytol 251, 41-77.

Harris, S. D. (2001). Septum formation in *Aspergillus nidulans*. Curr Opin Microbiol 4, 736-739.

Harris, S. D. and Momany, M. (2004). Polarity in filamentous fungi: moving beyond the yeast paradigm. Fungal Genet Biol 41, 391-400.

Helliwell, S. B., Howald, I., Barbet, N. and Hall, M. N. (1998). TOR2 is part of two related signaling pathways coordinating cell growth in *Saccharomyces cerevisiae*. Genetics 148, 99-112.

Hergovich, A., Stegert, M. R., Schmitz, D. and Hemmings, B. A. (2006). NDR kinases regulate essential cell processes from yeast to humans. Nat Rev Mol Cell Biol 7, 253-264.

Hirata, D., Kishimoto, N., Suda, M., Sogabe, Y., Nakagawa, S., Yoshida, Y., Sakai, K., Mizunuma, M., Miyakawa, T., Ishiguro, J. *et al.* (2002). Fission yeast Mor2/Cps12, a protein similar to *Drosophila* Furry, is essential for cell morphogenesis and its mutation induces Wee1-dependent G(2) delay. EMBO J 21, 4863-4874.

Ho, Y., Gruhler, A., Heilbut, A., Bader, G. D., Moore, L., Adams, S. L., Millar, A., Taylor, P., Bennett, K., Boutilier, K. *et al.* (2002). Systematic identification of protein complexes in *Saccharomyces cerevisiae* by mass spectrometry. Nature 415, 180-183.

Imai, J., Toh-e, A. and Matsui, Y. (1996). Genetic analysis of the *Saccharomyces cerevisiae* RHO3 gene, encoding a rho-type small GTPase, provides evidence for a role in bud formation. Genetics 142, 359-369.

Irazoqui, J. E., Gladfelter, A. S. and Lew, D. J. (2003). Scaffold-mediated symmetry breaking by Cdc42p. Nat Cell Biol 5, 1062-1070.

Ito, T., Chiba, T., Ozawa, R., Yoshida, M., Hattori, M. and Sakaki, Y. (2001). A comprehensive two hybrid analysis to explore the yeast protein interactome. Proc Natl Acad Sci U S A 98, 4569-4574.

Jaffe, A. B. and Hall, A. (2005). Rho GTPases: biochemistry and biology. Annu Rev Cell Dev Biol 21, 247-269.

Jorgensen, P., Nelson, B., Robinson, M. D., Chen, Y., Andrews, B., Tyers, M. and Boone, C. (2002).

High-resolution genetic mapping with ordered arrays of *Saccharomyces cerevisiae* deletion mutants. Genetics 162, 1091-1099.

Justice, R. W., Zilian, O., Woods, D. F., Noll, M. and Bryant, P. J. (1995). The *Drosophila* tumor suppressor gene warts encodes a homolog of human myotonic dystrophy kinase and is required for the control of cell shape and proliferation. Genes Dev 9, 534-546.

Kadrmas, J. L. and Beckerle, M. C. (2004). The LIM domain: from the cytoskeleton to the nucleus. Nat Rev Mol Cell Biol 5, 920-931.

Kagami, M., Toh-e, A. and Matsui, Y. (1997). SRO9, a multicopy suppressor of the bud growth defect in the *Saccharomyces cerevisiae rho3*-deficient cells, shows strong genetic interactions with tropomyosin genes, suggesting its role in organization of the actin cytoskeleton. Genetics 147, 1003- 1016.

Kanai, M., Kume, K., Miyahara, K., Sakai, K., Nakamura, K., Leonhard, K., Wiley, D. J., Verde, F., Toda, T. and Hirata, D. (2005). Fission yeast MO25 protein is localized at SPB and septum and is essential for cell morphogenesis. EMBO J 24, 3012-3025.

Katayama, S., Hirata, D., Arellano, M., Perez, P. and Toda, T. (1999). Fission yeast alpha-glucan synthase Mok1 requires the actin cytoskeleton to localize the sites of growth and plays an essential role in cell morphogenesis downstream of protein kinase C function. J Cell Biol 144, 1173-1186.

Konzack, S., Rischitor, P. E., Enke, C. and Fischer, R. (2005). The role of the kinesin motor KipA in microtubule organization and polarized growth of *Aspergillus nidulans*. Mol Biol Cell 16, 497-506.

Kubler, E. and Riezman, H. (1993). Actin and fimbrin are required for the internalization step of endocytosis in yeast. EMBO J 12, 2855-2862.

Latge, J. P. (1999). *Aspergillus fumigatus* and aspergillosis. Clin Microbiol Rev 12, 310-350.

Latge, J. P. (2007). The cell wall: a carbohydrate armour for the fungal cell. Mol Microbiol 66, 279-290.

Leonhard, K. and Nurse, P. (2005). Ste20/GCK kinase Nak1/Orb3 polarizes the actin cytoskeleton in fission yeast during the cell cycle. J Cell Sci 118, 1033-1044.

Levin, D. E. (2005). Cell wall integrity signaling in *Saccharomyces cerevisiae*. Microbiol Mol Biol Rev 69, 262-291.

Li, R., Zhang, B. and Zheng, Y. (1997). Structural determinants required for the interaction between Rho GTPase and the GTPase-activating domain of p190. J Biol Chem 272, 32830-32835.

Li, S., Du, L., Yuen, G. and Harris, S. D. (2006). Distinct ceramide synthases regulate polarized growth in the filamentous fungus *Aspergillus nidulans*. Mol Biol Cell 17, 1218-1227.

Lin, J. L., Chen, H. C., Fang, H. I., Robinson, D., Kung, H. J. and Shih, H. M. (2001). MST4, a new Ste20-related kinase that mediates cell growth and transformation via modulating ERK pathway. Oncogene 20, 6559-6569.

Lorberg, A., Schmitz, H. P., Jacoby, J. J. and Heinisch, J. J. (2001). Lrg1p functions as a putative GTPase-activating protein in the Pkc1p-mediated cell integrity pathway in *Saccharomyces cerevisiae*. Mol Genet Genomics 266, 514-526.

Madaule, P., Axel, R. and Myers, A. M. (1987). Characterization of two members of the *rho* gene family from the yeast *Saccharomyces cerevisiae*. Proc Natl Acad Sci U S A 84, 779-783.

Magan, N. (2007). Fungi in Extreme Environments. In: The Mycota IV: Environmental and Microbial Relationships, 85-104. K. Esser, C. P. Kubicek and I. S. Druzhinina (eds.), Springer-Verlag Berlin Heidelberg, Berlin.

Mahlert, M., Leveleki, L., Hlubek, A., Sandrock, B. and Bolker, M. (2006). Rac1 and Cdc42 regulate hyphal growth and cytokinesis in the dimorphic fungus *Ustilago maydis*. Mol Microbiol 59, 567-578.

Malavazi, I., Semighini, C. P., Kress, M. R., Harris, S. D. and Goldman, G. H. (2006). Regulation of hyphal morphogenesis and the DNA damage response by the *Aspergillus nidulans* ATM homolog AtmA. Genetics 173, 99-109.

Matsui, Y. and Toh-E, A. (1992a). Isolation and characterization of two novel ras superfamily genes in *Saccharomyces cerevisiae*. Gene 114, 43-49.

Matsui, Y. and Toh-E, A. (1992b). Yeast *RHO3* and *RHO4 ras* superfamily genes are necessary for bud growth, and their defect is suppressed by a high dose of bud formation genes *CDC42* and *BEM1*. Mol Cell Biol 12, 5690-5699.

Mazur, P. and Baginsky, W. (1996). *In vitro* activity of 1,3-beta-D-glucan synthase requires the GTPbinding protein Rho1. J Biol Chem 271, 14604-14609.

Minke, P. F., Lee, I. H. and Plamann, M. (1999). Microscopic analysis of *Neurospora ropy* mutants defective in nuclear distribution. Fungal Genet Biol 28, 55-67.

Mishra, N. C. and Tatum, E. L. (1972). Effect of L-sorbose on polysaccharide synthetases of *Neurospora crassa*. Proc Natl Acad Sci U S A 69, 313-317.

Mlodzik, M. (2002). Planar cell polarization: do the same mechanisms regulate *Drosophila* tissue polarity and vertebrate gastrulation? Trends Genet 18, 564-571.

Momany, M. (2002). Polarity in filamentous fungi: establishment, maintenance and new axes. Curr Opin Microbiol 5, 580-585.

Morii, N., Kawano, K., Sekine, A., Yamada, T. and Narumiya, S. (1991). Purification of GTPase activating protein specific for the *rho* gene products. J Biol Chem 266, 7646-7650.

Morris, S. J., Friese, C. F. and Allen, M. F. (2007). Disturbance in Natural Ecosystems: Scaling from Fungal Diversity to Ecosystem Functioning. In: The Mycota IV: Environmental and Microbial Relationships, 31-46. K. Esser, C. P. Kubicek and I. S. Druzhinina (eds.), Springer-Verlag Berlin Heidelberg, Berlin.

Mulholland, J., Preuss, D., Moon, A., Wong, A., Drubin, D. and Botstein, D. (1994). Ultrastructure of the yeast actin cytoskeleton and its association with the plasma membrane. J Cell Biol 125, 381-391.

Nakano, K., Mutoh, T., Arai, R. and Mabuchi, I. (2003). The small GTPase Rho4 is involved in controlling cell morphology and septation in fission yeast. Genes Cells 8, 357-370.

Nakano, K., Mutoh, T. and Mabuchi, I. (2001). Characterization of GTPase-activating proteins for the function of the Rho-family small GTPases in the fission yeast *Schizosaccharomyces pombe*. Genes Cells 6, 1031-1042.

Nargang, F. E., Kunkele, K. P., Mayer, A., Ritzel, R. G., Neupert, W. and Lill, R. (1995). 'Sheltered disruption' of *Neurospora crassa* MOM22, an essential component of the mitochondrial protein import complex. EMBO J 14, 1099-1108.

Nelson, B., Kurischko, C., Horecka, J., Mody, M., Nair, P., Pratt, L., Zougman, A., McBroom, L. D., Hughes, T. R., Boone, C. *et al.* (2003). RAM: a conserved signaling network that regulates Ace2p transcriptional activity and polarized morphogenesis. Mol Biol Cell 14, 3782-3803.

Nelson, W. J. (2003). Adaptation of core mechanisms to generate cell polarity. Nature 422, 766-774.

Ogawa, K., Hosoya, H., Yokota, E., Kobayashi, T., Wakamatsu, Y., Ozato, K., Negishi, S. and Obika, M. (1987). Melanoma dynein: evidence that dynein is a general "motor" for microtubule-associated cell motilities. Eur J Cell Biol 43, 3-9.

Olofsson, B. (1999). Rho guanine dissociation inhibitors: pivotal molecules in cellular signalling. Cell Signal 11, 545-554.

Orbach, M. J. (1984). A cosmid with a HyR marker for fungal library construction and screening. Gene 150, 159-162.

Orr-Weaver, T. L. and Szostak, J. W. (1983). Yeast recombination: the association between doublestrand gap repair and crossing-over. Proc Natl Acad Sci U S A 80, 4417-4421.

Ozaki, K., Tanaka, K., Imamura, H., Hihara, T., Kameyama, T., Nonaka, H., Hirano, H., Matsuura, Y. and Takai, Y. (1996). Romlp and Rom2p are GDP/GTP exchange proteins (GEPs) for the Rho1p small GTP binding protein in *Saccharomyces cerevisiae*. EMBO J 15, 2196-2207.

Palanivelu, R. and Preuss, D. (2000). Pollen tube targeting and axon guidance: parallels in tip growth mechanisms. Trends Cell Biol 10, 517-524.

Palazzo, A. F., Cook, T. A., Alberts, A. S. and Gundersen, G. G. (2001). mDia mediates Rho-regulated formation and orientation of stable microtubules. Nat Cell Biol 3, 723-729.

Park, H. O. and Bi, E. (2007). Central roles of small GTPases in the development of cell polarity in yeast and beyond. Microbiol Mol Biol Rev 71, 48-96.

Plamann, M., Minke, P. F., Tinsley, J. H. and Bruno, K. S. (1994). Cytoplasmic dynein and actin-related protein Arp1 are required for normal nuclear distribution in filamentous fungi. J Cell Biol 127, 139-149.

Pöggeler, S., Masloff, S., Hoff, B., Mayrhofer, S. and Kück, U. (2003). Versatile EGFP reporter plasmids for cellular localization of recombinant gene products in filamentous fungi. Curr Genet 43, 54- 61.

Ponton, J., Ruchel, R., Clemons, K. V., Coleman, D. C., Grillot, R., Guarro, J., Aldebert, D., Ambroise-Thomas, P., Cano, J., Carrillo-Munoz, A. J. *et al.* (2000). Emerging pathogens. Med Mycol 38 Suppl 1, 225-236.

Pringle, A. and Taylor, J. (2002). The fitness of filamentous fungi. Trends Microbiol 10, 474-481. Pruyne, D. and Bretscher, A. (2000a). Polarization of cell growth in yeast. I. Establishment and maintenance of polarity states. J Cell Sci 113, 365-375.

Pruyne, D. and Bretscher, A. (2000b). Polarization of cell growth in yeast. II. The role of the cortical actin cytoskeleton. J Cell Sci 113, 571-585.

Pruyne, D., Legesse-Miller, A., Gao, L., Dong, Y. and Bretscher, A. (2004). Mechanisms of polarized growth and organelle segregation in yeast. Annu Rev Cell Dev Biol 20, 559-591.

Puig, O., Caspary, F., Rigaut, G., Rutz, B., Bouveret, E., Bragado-Nilsson, E., Wilm, M. and Seraphin, B. (2001). The tandem affinity purification (TAP) method: a general procedure of protein complex purification. Methods 24, 218-229.

Punt, P. J., van Biezen, N., Conesa, A., Albers, A., Mangnus, J. and van den Hondel, C. (2002).

Filamentous fungi as cell factories for heterologous protein production. Trends Biotechnol 20, 200-206.

Qadota, H., Python, C. P., Inoue, S. B., Arisawa, M., Anraku, Y., Zheng, Y., Watanabe, T., Levin, D. E. and Ohya, Y. (1996). Identification of yeast Rho1p GTPase as a regulatory subunit of 1,3-beta-glucan synthase. Science 272, 279-281.

Qian, Z., Lin, C., Espinosa, R., LeBeau, M. and Rosner, M. R. (2001). Cloning and characterization of MST4, a novel Ste20-like kinase. J Biol Chem 276, 22439-22445.

Racki, W. J., Becam, A. M., Nasr, F. and Herbert, C. J. (2000). Cbk1p, a protein similar to the human myotonic dystrophy kinase, is essential for normal morphogenesis in *Saccharomyces cerevisiae*. EMBO J 19, 4524-4532.

Rasmussen, C. G. and Glass, N. L. (2005). A Rho-type GTPase, *rho-4*, is required for septation in *Neurospora crassa*. Eukaryot Cell 4, 1913-1925.

Rasmussen, C. G. and Glass, N. L. (2007). Localization of RHO-4 Indicates Differential Regulation of Conidial versus Vegetative Septation in the Filamentous Fungus *Neurospora crassa*. Eukaryot Cell 6, 1097-1107.

Ridley, A. J. (2006). Rho GTPases and actin dynamics in membrane protrusions and vesicle trafficking. Trends Cell Biol 16, 522-529.

Ridley, A. J. (1995). Rho-related proteins: actin cytoskeleton and cell cycle. Curr Opin Genet Dev 5, 24- 30.

Rodal, A. A., Kozubowski, L., Goode, B. L., Drubin, D. G. and Hartwig, J. H. (2005). Actin and septin ultrastructures at the budding yeast cell cortex. Mol Biol Cell 16, 372-384.

Rossman, K. L., Worthylake, D. K., Snyder, J. T., Siderovski, D. P., Campbell, S. L. and Sondek, J. (2002). A crystallographic view of interactions between Dbs and Cdc42: PH domain-assisted guanine nucleotide exchange. EMBO J 21, 1315-1326.

Rottmann, M., Dieter, S., Brunner, H. and Rupp, S. (2003). A screen in *Saccharomyces cerevisiae* identified *CaMCM1*, an essential gene in *Candida albicans* crucial for morphogenesis. Mol Microbiol 47, 943-959.

Roumanie, O., Peypouquet, M. F., Thoraval, D., Doignon, F. and Crouzet, M. (2002). Functional interactions between the *VRP1-LAS17* and *RHO3-RHO4* genes involved in actin cytoskeleton organization in *Saccharomyces cerevisiae*. Curr Genet 40, 317-325.

Roumanie, O., Weinachter, C., Larrieu, I., Crouzet, M. and Doignon, F. (2001). Functional characterization of the Bag7, Lrgl and Rgd2 RhoGAP proteins from *Saccharomyces cerevisiae*. FEBS Lett 506, 149-156.

Sambrook, J., Fritsch, E. F. and Maniatis, T. (eds.) (1989). Molecular Cloning: A Laboratory Manual. Cold Spring Harbor Laboratory Press, New York.

Santos, B., Gutierrez, J., Calonge, T. M. and Perez, P. (2003). Novel Rho GTPase involved in cytokinesis and cell wall integrity in the fission yeast *Schizosaccharomyces pombe*. Eukaryot Cell 2, 521-533.

Schaller, M. D. (2001). Paxillin: a focal adhesion-associated adaptor protein. Oncogene 20, 6459-6472.

Schmeichel, K. L. and Beckerle, M. C. (1997). Molecular dissection of a LIM domain. Mol Biol Cell 8, 219-230.

Schmidt, A. and Hall, A. (2002). Guanine nucleotide exchange factors for Rho GTPases: turning on the switch. Genes Dev 16, 1587-1609.

Schneper, L., Krauss, A., Miyamoto, R., Fang, S. and Broach, J. R. (2004). The Ras/protein kinase A pathway acts in parallel with the Mob2/Cbk1 pathway to effect cell cycle progression and proper bud site selection. Eukaryot Cell 3, 108-120.

Schuchardt, I., Assmann, D., Thines, E., Schuberth, C. and Steinberg, G. (2005). Myosin-V, Kinesin-1, and Kinesin-3 cooperate in hyphal growth of the fungus *Ustilago maydis*. Mol Biol Cell 16, 5191-5201.

Seiler, S., Nargang, F. E., Steinberg, G. and Schliwa, M. (1997). Kinesin is essential for cell morphogenesis and polarized secretion in *Neurospora crassa*. EMBO J 16, 3025-3034.

Seiler, S. and Plamann, M. (2003). The genetic basis of cellular morphogenesis in the filamentous fungus *Neurospora crassa*. Mol Biol Cell 14, 4352-4364.

Seiler, S., Plamann, M. and Schliwa, M. (1999). Kinesin and dynein mutants provide novel insights into the roles of vesicle traffic during cell morphogenesis in *Neurospora*. Curr Biol 9, 779-785.

Seiler, S., Vogt, N., Ziv, C., Gorovits, R. and Yarden, O. (2006). The STE20/germinal center kinase POD6 interacts with the NDR kinase COT1 and is involved in polar tip extension in *Neurospora crassa*. Mol Biol Cell 17, 4080-4092.

Sharpless, K. E. and Harris, S. D. (2002). Functional characterization and localization of the *Aspergillus nidulans* formin SEPA. Mol Biol Cell 13, 469-479.

Sikorski, R. S. and Hieter, P. (1989). A system of shuttle vectors and yeast host strains designed for efficient manipulation of DNA in *Saccharomyces cerevisiae*. Genetics 122, 19-27.

Snyder, J. T., Worthylake, D. K., Rossman, K. L., Betts, L., Pruitt, W. M., Siderovski, D. P., Der, C. J. and Sondek, J. (2002). Structural basis for the selective activation of Rho GTPases by Dbl exchange factors. Nat Struct Biol 9, 468-475.

Springer, M. L. and Yanofsky, C. (1989). A morphological and genetic analysis of conidiophore development in *Neurospora crassa*. Genes Dev 3, 559-571.

Stegert, M. R., Hergovich, A., Tamaskovic, R., Bichsel, S. J. and Hemmings, B. A. (2005). Regulation of NDR protein kinase by hydrophobic motif phosphorylation mediated by the mammalian Ste20-like kinase MST3. Mol Cell Biol 25, 11019-11029.

Steinberg, G. (2007). Hyphal growth: a tale of motors, lipids, and the Spitzenkörper. Eukaryot Cell 6, 351-360.

Steinberg, G. (2000). The cellular roles of molecular motors in fungi. Trends Microbiol 8, 162-168. Stewart, M. S., Krause, S. A., McGhie, J. and Gray, J. V. (2007). Mpt5p, a stress tolerance- and lifespan-promoting PUF protein in *Saccharomyces cerevisiae*, acts upstream of the cell wall integrity pathway. Eukaryot Cell 6, 262-270.

Sussman, A., Huss, K., Chio, L. C., Heidler, S., Shaw, M., Ma, D., Zhu, G., Campbell, R. M., Park, T. S., Kulanthaivel, P. *et al.* (2004). Discovery of cercosporamide, a known antifungal natural product, as a selective Pkc1 kinase inhibitor through high-throughput screening. Eukaryot Cell 3, 932-943.

Taft, C. S., Zugel, M. and Selitrennikoff, C. P. (1991). *In vitro* inhibition of stable 1,3-beta-D-glucan synthase activity from *Neurospora crassa*. J Enzyme Inhib 5, 41-49.

Tamaskovic, R., Bichsel, S. J. and Hemmings, B. A. (2003). NDR family of AGC kinases – essential regulators of the cell cycle and morphogenesis. FEBS Lett 546, 73-80.

Tcherkezian, J. and Lamarche-Vane, N. (2007). Current knowledge of the large RhoGAP family of proteins. Biol Cell 99, 67-86.

Terenzi, H. F. and Reissig, J. L. (1967). Modifiers of the *cot* gene in *Neurospora*: the gulliver mutants. Genetics 56, 321-329.

Tu, S. S., Wu, W. J., Yang, W., Nolbant, P., Hahn, K. and Cerione, R. A. (2002). Antiapoptotic Cdc42 mutants are potent activators of cellular transformation. Biochemistry 41, 12350-12358.

Utsugi, T., Minemura, M., Hirata, A., Abe, M., Watanabe, D. and Ohya, Y. (2002). Movement of yeast 1,3-beta-glucan synthase is essential for uniform cell wall synthesis. Genes Cells 7, 1-9.

Valdivia, R. H., Baggott, D., Chuang, J. S. and Schekman, R. W. (2002). The yeast clathrin adaptor protein complex 1 is required for the efficient retention of a subset of late Golgi membrane proteins. Dev Cell 2, 283-294.

Valdivia, R. H. and Schekman, R. (2003). The yeasts Rho1p and Pkc1p regulate the transport of chitin synthase III (Chs3p) from internal stores to the plasma membrane. Proc Natl Acad Sci U S A 100, 10287-10292.

Vale, R. D. (2003). The molecular motor toolbox for intracellular transport. Cell 112, 467-480. Vale, R. D., Reese, T. S. and Sheetz, M. P. (1985). Identification of a novel force-generating protein, kinesin, involved in microtubule-based motility. Cell 42, 39-50.

Van Aelst, L. and D'Souza-Schorey, C. (1997). Rho GTPases and signaling networks. Genes Dev 11, 2295-2322.

Vanni, C., Ottaviano, C., Guo, F., Puppo, M., Varesio, L., Zheng, Y. and Eva, A. (2005). Constitutively active Cdc42 mutant confers growth disadvantage in cell transformation. Cell Cycle 4, 1675-1682.

Varelas, X., Stuart, D., Ellison, M. J. and Ptak, C. (2006). The Cdc34/SCF ubiquitination complex mediates *Saccharomyces cerevisiae* cell wall integrity. Genetics 174, 1825-1839.

Verde, F., Wiley, D. J. and Nurse, P. (1998). Fission yeast *orb6*, a ser/thr protein kinase related to mammalian rho kinase and myotonic dystrophy kinase, is required for maintenance of cell polarity and coordinates cell morphogenesis with the cell cycle. Proc Natl Acad Sci U S A 95, 7526-7531.

Versele, M. and Thevelein, J. M. (2001). Lre1 affects chitinase expression, trehalose accumulation and heat resistance through inhibition of the Cbk1 protein kinase in *Saccharomyces cerevisiae*. Mol Microbiol 41, 1311-1326.

Virag, A., Lee, M. P., Si, H. and Harris, S. D. (2007). Regulation of hyphal morphogenesis by *cdc42* and *rac1* homologues in *Aspergillus nidulans*. Mol Microbiol 66, 1579-1596.

Vollmer, S. J. and Yanofsky, C. (1986). Efficient cloning of genes of *Neurospora crassa*. Proc Natl Acad Sci U S A 83, 4869-4873.

Walsh, T. J., Groll, A., Hiemenz, J., Fleming, R., Roilides, E. and Anaissie, E. (2004). Infections due to emerging and uncommon medically important fungal pathogens. Clin Microbiol Infect 10 Suppl 1, 48- 66.

Watabe-Uchida, M., Govek, E. E. and Van Aelst, L. (2006). Regulators of Rho GTPases in neuronal development. J Neurosci 26, 10633-10635.

Watanabe, D., Abe, M. and Ohya, Y. (2001). Yeast Lrg1p acts as a specialized RhoGAP regulating 1,3-beta-glucan synthesis. Yeast 18, 943-951.

Weinzierl, G., Leveleki, L., Hassel, A., Kost, G., Wanner, G. and Bolker, M. (2002). Regulation of cell separation in the dimorphic fungus *Ustilago maydis*. Mol Microbiol 45, 219-231.

Weiss, E. L., Kurischko, C., Zhang, C., Shokat, K., Drubin, D. G. and Luca, F. C. (2002). The *Saccharomyces cerevisiae* Mob2p-Cbk1p kinase complex promotes polarized growth and acts with the mitotic exit network to facilitate daughter cell-specific localization of Ace2p transcription factor. J Cell Biol 158, 885-900.

Wendland, J. (2001). Comparison of morphogenetic networks of filamentous fungi and yeast. Fungal Genet Biol 34, 63-82.

Wendland, J. and Philippsen, P. (2001). Cell polarity and hyphal morphogenesis are controlled by multiple rho-protein modules in the filamentous ascomycete *Ashbya gossypii*. Genetics 157, 601-610.

Wendland, J. and Walther, A. (2006). Septation and Cytokinesis in Fungi. In: The Mycota I: Growth, Differentiation and Sexuality, 105-122. K. Esser, U. Kües and R. Fischer (eds.), Springer-Verlag Berlin Heidelberg, Berlin.

Wennerberg, K. and Der, C. J. (2004). Rho-family GTPases: it's not only Rac and Rho (and I like it). J Cell Sci 117, 1301-1312.

Wiley, D. J., Marcus, S., D'Urso, G. and Verde, F. (2003). Control of cell polarity in fission yeast by association of Orb6p kinase with the highly conserved protein methyltransferase Skb1p. J Biol Chem 278, 25256-25263.

6 Supplementary data

6.1 Lrg1p and Cbk1p are involved in pseudohyphae formation in *Saccharomyces cerevisiae*

Polar growth defects of deletion mutations of *S. cerevisiae* orthologous of *lrg-1* and *cot-1* were investigated. Therefore the strains Y33937 (Δ*Lrg1*), Y22051 (Δ*Cbk1*, the *cot-1* homologue) and BY4743 (*wild type*) were grown on SLAD-media that induces pseudohyphal formation. BY4743 shows high amounts of differentiated pseudohyphal growth after 5 days of incubation. In contrast, much less pseudohyphae formation was observed in Y33937 and Y22051 (Figure 21), indicating a function for the conserved proteins identified in *N. crassa* also in *S. cerevisiae* in polar morphogenesis, in yeast as pseudohyphal growth. Nevertheless, the formation of pseudohyphae still takes place in few colonies under these conditions, indicating that the functions of the proteins are not essential for pseudohyphal formation.

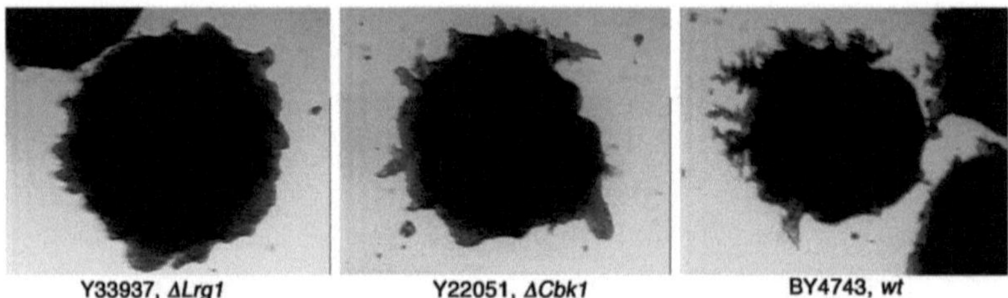

Figure 21: *Lrg1* and *Cbk1* are necessary for normal pseudohyphal formation in *Saccharomyces cerevisiae*.
Strains were grown on SLAD Medium to induce pseudohyphae formation for 5 days. Representative single colonies are shown. Pseudohyphal growth is strongly inhibited in Δ*Cbk1* and nearly absent in Δ*Lrg1* strains.

6.2 CDC24 enhances GDP-GTP exchange *in vitro*

Morphological differences in several conditional *cdc24* alleles that are defective in polar elongation suggest, that CDC24 may regulate two related Rho proteins in *N. crassa* (Stephan Seiler, unpublished observations). One of them is CDC42, which is activated by CDC24 in *S. cerevisiae*. The other Rho protein is RAC, which is absent in yeasts. To elevate, if RAC and CDC42 are targets of CDC24, fluorescent *in vitro* studies of Rho proteins, preloaded with Mant-GDP for GEF activity were performed (initially in cooperation with Jan Faix, Hannover medical school). The impact of the plecstrin homology (PH) domain on the interaction and regulation of guanidine nucleotide exchange activity of the corresponding Rho GTPase for DH (Dbl-homologous)-PH tandem containing GEFs was previously shown (Rossman *et al.*, 2002; Snyder *et al.*, 2002). Therefore, the DH and PH domains of CDC24 (amino acids 204-544) were heterologously purified as maltose binding protein (MBP) fusion proteins (encoded from pNV85) from *E. coli* and used to determine the biochemical GEF activity of CDC24 towards all *N. crassa* Rho GTPases. Increased fluorescence enhancement, which correlates with increased GTPase exchange activity, was specifically detected for CDC42 and RAC (Figure 22). These results fit well with the genetic and morphological data.

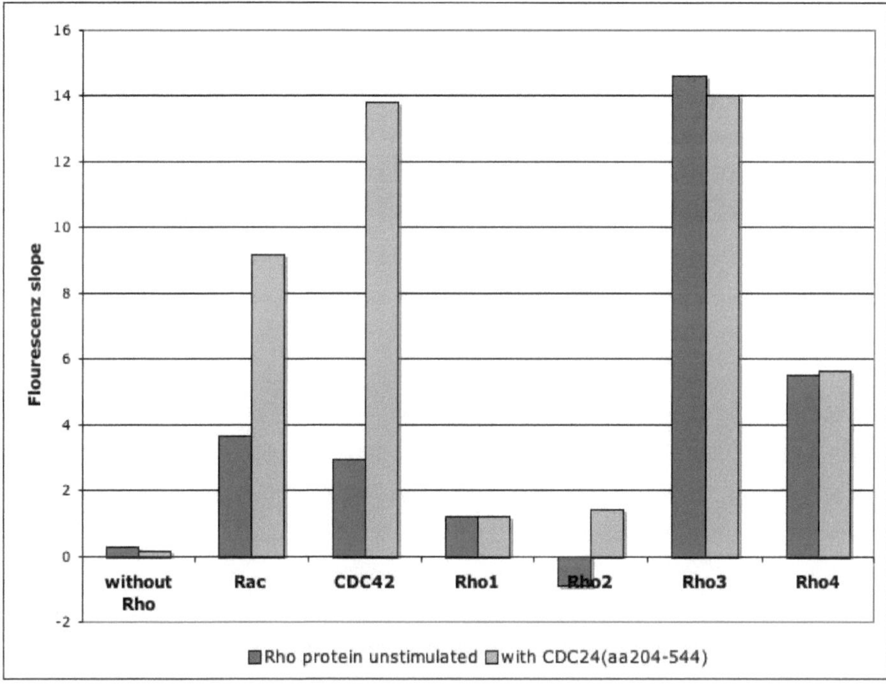

Figure 22: CDC24 regulates CDC42 and RAC *in vitro*.
Fluorescence intensity of the six Rho proteins in presence of Mant-GDP was measured all 20 seconds for 10 min and the slope in intensity/min is shown. RAC and CDC42 show a stimulation of their exchange activity.

6.3 Vector information

6.3.1 Cloning plasmid pNV86

The subcloning vector pNV86 or pPiNatZ (pPicholi-C, containing nourseothricin (*nat*) and zeocin resistances) was generated. The vector can be used in *E. coli* and *Pichia pastoris* and contains a zeocin selection for these two organisms. When ampicillin or kanamycin resistant vectors are used in further cloning steps, there are no more background colonies resulting from this subcloning vector. To achieve pNV86, the pPicholi-C (MoBiTec, Göttingen, Germany) backbone was in a first step cut with *Eco*32I in the corresponding buffer ON, followed by cleavage with mung bean nuclease after adding 0.5 mM $ZnCl_2$ to the buffer for 15 min at room temperature. The vector was purified via agarose gel electrophorese and subsequently religated in the *Eco32*I restriction buffer containing ATP. A second cut with *Eco32*I reduces the amount of transformants still containing the cut site. The plasmid derived from transformants were analysed to ensure the deletion of the restriction site of the obtained pPicholi-CΔ*Eco*RV. An analogous procedure was used to delete the *Nco*I site between the promoter and the nourseothricin resistance in the plasmid pNV1, leading to plasmid pNV63, which was further tested for mediation of the nourseothricin resistance to *N. crassa*. To derive the subcloning vector, the promoter P_{lacUV5} and the following suicide gene encoding the *Eco47*IR restriction enzyme were derived from pJet1. To enable the cloning of this suicide gene, the GFP encoding gene amplified from pSM1 (Pöggeler *et al.*, 2003) via primers NV_GFP1 and NV_GFP2 was inserted as *Eco32*I fragment into the pJet1 vector, resulting in pNV5 (alternative name pJet-G). The cloning vector pPiNatZ was obtained by cutting the pPicholi-CΔ*Eco*RV with *Not*I and *Sac*I and introducing the silenced suicide gene from pJet-G as *Sal*I and *Bsp120*I (in *Not*I) fragment together with P_{gpdA}*::nat*^{R} gene from pNV5 as *Sac*I and *Xho*I (in *Sal*I).
The obtained pNV86 cloning vector was cut with *Eco32*I to get rid of the GFP fragment for cloning purposes. It contains dominant selection for fragment insertion, a zeocin, but no ampicillin or kanamycin resistance for *E*. coli and *P. pastoris,* and a nourseothricin resistance under control of *Aspergillus gpdA* promoter for direct transformation in *N. crassa* (Figure 23).

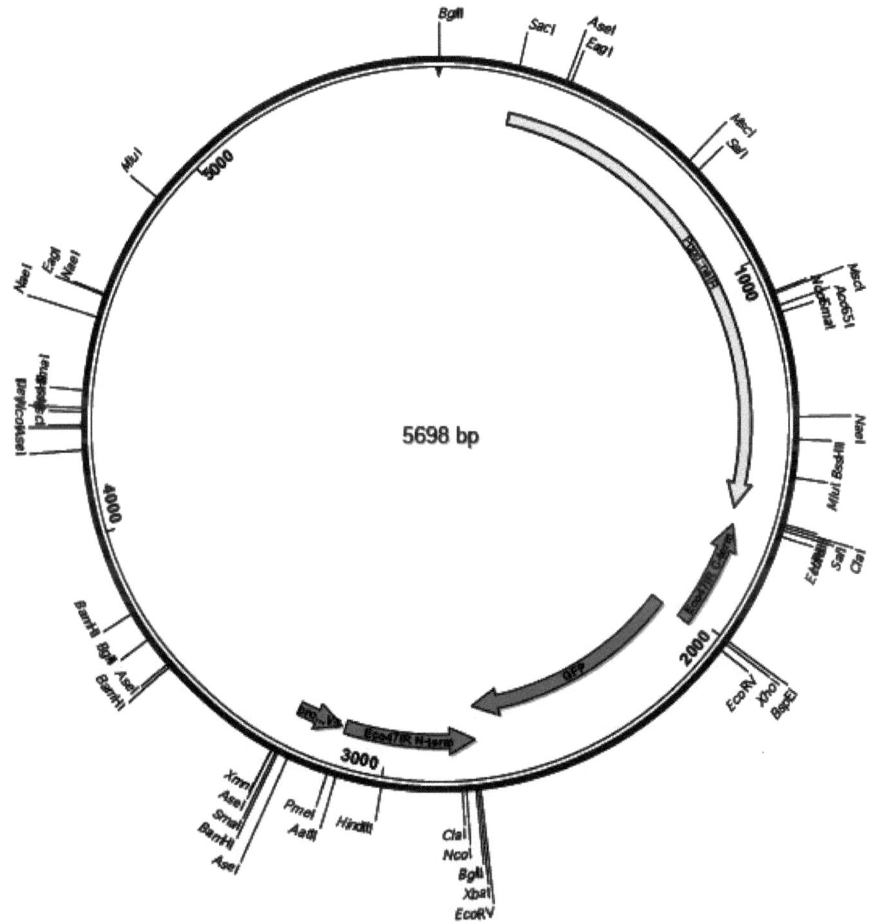

Figure 23: Subcloning vector pNV86
The backbone is derived from pPicholi-C, were the P_{CUP1} is replaced with the nourseothricin resistance and by the *GFP* interrupted suicide gene *Eco47*IR that are indicated by yellow and red boxes, respectively.

6.3.2 Expression plasmids pNV87 and pNV88

For expression of large eukaryotic proteins like the full length CDC24 from *N. crassa*, the pFastBac Dual (Invitrogen, Carlsbad, USA) was modified with a hexahistidine epitop that is recognised by specific antibodies (RGS-His^6) and a GST tag for purification under control of the *PH* promoter.

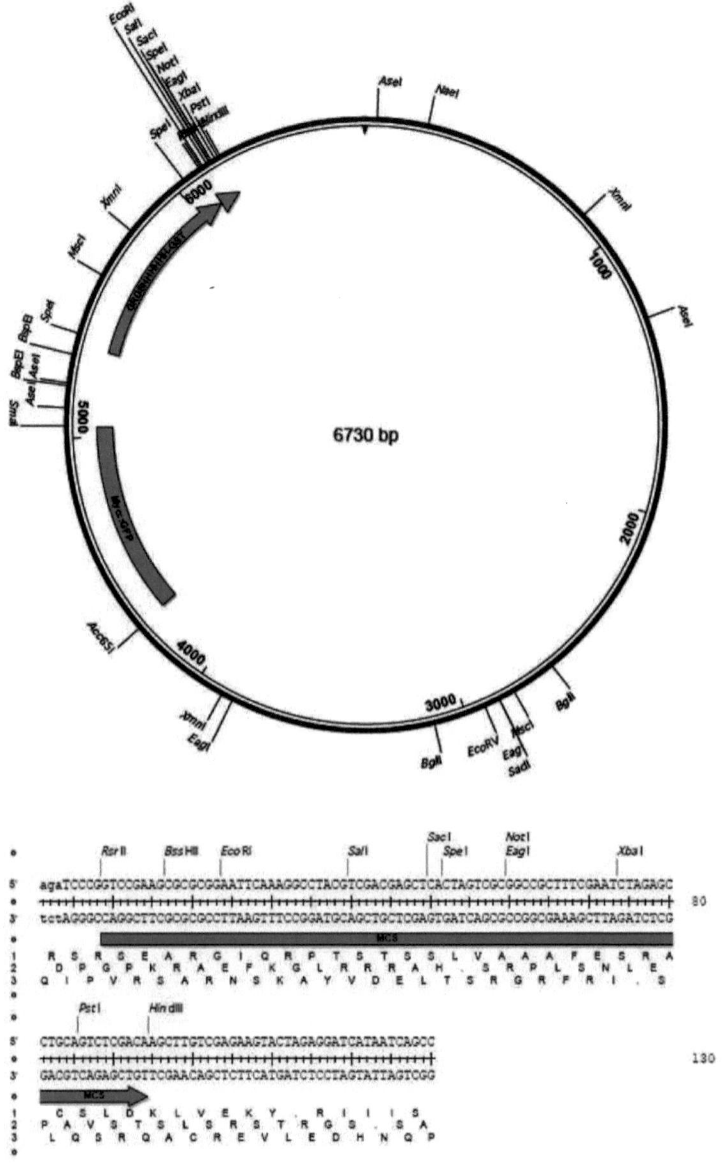

Figure 24: Map and multiple cloning site of pNV88

The map (upper panel) shows the encoded MYC::GFP and GRGSHHHHHH::GST modifications of the pFastBac Dual vector (red boxes). The multiple cloning site is shown in the lower panel.

Therefore, the GST epitop was amplified with the sense primer NV_GST2 that inserted the MGRGSHHHHHH coding sequence and the reverse primer NV_GST3 and ligated via the inserted *Bgl*II site in the *Bam*HI site of pFastBac Dual to obtain pNV87. For fast detection of the transfection efficiency of the SF9 cells, GFP was amplified with NV_GFP5 and NV_GFP10 thereby adding a MYC tag N-terminally under the control of the *P10* promoter and inserted as *Bgl*II/*Acc65*I fragment in *Bpi*I/*Acc65*I of pNV87 to obtain pNV88 (Figure 24).

6.3.3 Modification of the multiple cloning sites of pMal-c2x to obtain pNV72

To obtain the reading frame of pPicholi-C, which was initially used to express the RHO proteins (data not shown), the multiple cloning sites (MCS) of pMal-c2x was changed. The modification of the MCS in pMal-c2x to obtain pNV72 was inserted with a PCR using primers NV_link7 and NV_MBP1 with pMal-c2x as template. The amplicon containing of the changed multiple cloning site and a part of the MalE coding sequence was cut with *Bgl*II and *Hind*III and ligated into pMal-c2x cut with the same restriction enzymes to obtain pNV72. Figure 25 shows the changed MCS of pNV72.

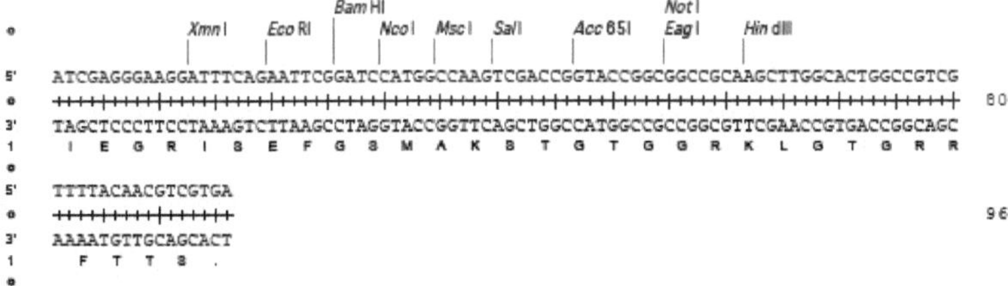

Figure 25: Changed multiple cloning site of the modified pMal-c2x.
The MCS of pMal-c2x was adapted to pPicholi-C. Sequence is shown until the stop codon.

6.4 Searching for interacting proteins of LRG1: tandem affinity purification

To address the question, which protein might interact with LRG1, a protein complex purification was attempt. Therefore, the tandem affinity purification (TAP) coding sequence was amplified with the oligonucleotides NV_tap_n_r and NV_tap_N_1 from pBS1761 (Puig *et al.*, 2001) and was ligated as *Nco*I/*Eco32*I fragment together with the *lrg-1* promoter amplified with LRG-5SacI and LRG-3NcoI as *Sac*I/*Nco*I fragment into *Sac*I/*Msc*I cut pNV16. The hygromycin cassette was integrated as *Bsp120*I fragment into the *Not*I site to obtain pNV75.

The purification procedure did not result in the expected pure complex and therefore the single steps were analysed in more detail (Figure 26 A). The α-LRG1 antiserum detected in a dilution of

1:100000 a single band in western blots of *wild type* extracts, and a second band occurs at higher molecular weight for extracts of the *tap::lrg-1* strain. This tagged protein could be enriched and a similar recognition by a α-LRG1 antiserum and an antibody against protein A indicates a specific detection of the TAP::LRG1 fusion protein. In the case of TAP::LRG1, the fusion protein was not cleaved by TEV protease, whereas a control substrate (kindly provided by Dr. Alf Herzig, Max Planck Institute for biophysical chemistry, Göttingen, Germany) could be efficiently cleaved under the same buffer conditions (Figure 26 B).

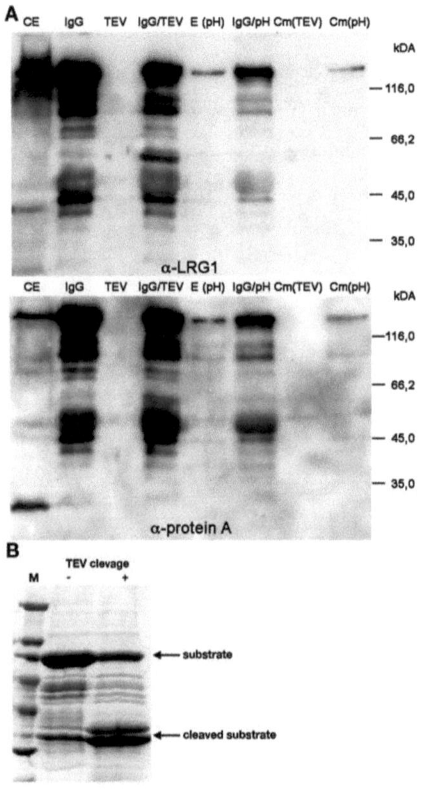

Figure 26: TAP::LRG1 is not cleavable by the TEV protease.
(A) Ectopically expressed TAP::LRG1, which rescues *lrg-1(12-20)* growth defects, was detected by the LRG1 antiserum (upper panel) and with a-protein A antibody (lower panel) through different purification steps. CE: crude extract, IgG: protein bound to IgG sepherose, TEV: elution fraction cleaved with TEV protease, E (pH): elution by glycin buffer, pH 1.8 from IgG sepharose, IgG/pH: IgG beads after the pH elution, Cm(TEV): calmodulin beads loaded with TEV eluate from IgG, Cm (pH): calmodulin beads loaded with fast neutralized eluate from pH elution of IgG bound proteins. (B) The TEV protease was able to cleave a control substrate incubated in the same buffer as the TAP::LRG1 crude extract at 4°C for 4 hours.

Printed by Books on Demand GmbH, Norderstedt / Germany